T0092982

A MOST INTERESTING PROBLEM

FRONTISPIECE. Charles Darwin in 1871. (Photograph by Oscar Gustaf Rejlander. Courtesy of the Cambridge University Library)

A Most Interesting Problem

WHAT DARWIN'S *DESCENT OF MAN*
GOT RIGHT AND WRONG ABOUT
HUMAN EVOLUTION

EDITED BY
JEREMY M. DESILVA

WITH AN INTRODUCTION
BY JANET BROWNE

PRINCETON UNIVERSITY PRESS
PRINCETON & OXFORD

Published by Princeton University Press
41 William Street, Princeton, New Jersey 08540
6 Oxford Street, Woodstock, Oxfordshire OX20 1TR

press.princeton.edu

All Rights Reserved

Library of Congress Cataloging-in-Publication Data
Names: DeSilva, Jeremy M., 1976- editor.
Title: A most interesting problem : what Darwin's Descent of man got right
    and wrong about human evolution / edited by Jeremy M. DeSilva.
Description: Princeton : Princeton University Press, [2021] | Includes
    bibliographical references and index.
Identifiers: LCCN 2020012362 (print) | LCCN 2020012363 (ebook) | ISBN
    9780691191140 (hardback) | ISBN 9780691210810 (ebook)
Subjects: LCSH: Darwin, Charles, 1809-1882. Descent of man. | Sexual
    selection in animals. | Sexual dimorphism (Animals) | Sex differences. |
    Human beings—Origin.
Classification: LCC QH365.D8 M67 2021 (print) | LCC QH365.D8 (ebook) |
    DDC 591.56/2—dc23
LC record available at https://lccn.loc.gov/2020012362
LC ebook record available at https://lccn.loc.gov/2020012363

British Library Cataloging-in-Publication Data is available

Editorial: Alison Kalett and Abigail Johnson
Production Editorial: Ellen Foos
Text Design: Leslie Flis
Jacket Design: Pamela Schnitter
Production: Jacqueline Poirier
Publicity: Sarah Henning-Stout and Katie Lewis
Copyeditor: Amy K. Hughes

Jacket Art: Shutterstock

This book has been composed in Arno Pro

Printed on acid-free paper. ∞

Printed in the United States of America

10 9 8 7 6 5 4 3 2 1

To Charles, and the questions he dared to ask

CONTENTS

LIST OF ILLUSTRATIONS

PREFACE

ON A COLD AND DRIZZLY February afternoon, I walked from bustling Euston Station to 4 Chester Place Road, a luxurious, cream-colored three-story, five-bedroom flat with large windows facing west toward the greenery of one of London's royal parks. I didn't knock, knowing of course that Sarah, Emma, or Charles would not be there to answer. For several minutes, I stared up at the large second-floor windows, trying to imagine the scene a century and a half earlier when Charles Darwin held in his hands, for the first and only time, the ancient fossilized skull of an extinct human.

From August 25 to September 1, 1864, the resident of 4 Chester Place Road, Sarah Elizabeth Wedgwood, hosted her younger sister Emma Darwin and Emma's husband, Charles. Charles had been quite ill, and the Darwins had traveled from their country home at Downe, in Kent, to stay for a week to, as Charles wrote, "see how I stand a change."[1]

The Wedgwood flat was in the perfect location for Charles. It was about a kilometer from the home of geologist Charles Lyell, a close friend and colleague of Darwin's, and within walking distance to the botanical gardens maintained by the Royal Botanical Society and the Zoological Society of London. It would be a good place for Charles to rest his body while keeping his mind active as he finished his manuscript on climbing plants. As he wrote at the end of 1864, "I have suffered from almost incessant vomiting for nine months, & that has so weakened my brain, that any excitement brings on whizzing & fainting feelings."[2]

The unusual fossilized skull brought to Darwin in the summer of 1864 had been discovered in Forbes Quarry, Gibraltar, in 1848 (see Preface figure). But it wasn't given much attention by the scientific community for over a decade. In September of 1864, George Busk, an English paleontologist, and Hugh Falconer, a Scottish paleontologist, arranged to have the skull displayed at the meeting of the British Association for the Advancement of Science in Bath to complement the papers they would deliver on the fossils from Gibraltar. But prior to the meeting, Falconer brought the Gibraltar skull to 4 Chester Place Road, London. There, it could be examined by his friend Charles Darwin.[3]

What is remarkable about this event is just how small an impact this fossil skull had on Darwin. The only evidence that this meeting occurred at all is a throwaway line in a September 1, 1864, letter Darwin wrote to Joseph Dalton Hooker when he returned to Down House. "Both Lyell & Falconer called on me & I was very glad to see them. F. brought me the wonderful Gibralter [sic] skull.—Farewell. Ever Yours | C. Darwin,"[4] he wrote to his good friend on a Thursday evening. Falconer himself recorded nothing about the meeting and died just five months later. If Darwin made sketches of the skull or jotted down any notes, they are lost.

As a paleoanthropologist who studies the human fossil record, I find this unsettling. How could the great Charles Darwin hold this skull— recognized today as a female Neanderthal—and not see, with his legendary observational skills, the significance of it?

As I looked up into the windows of 4 Chester Place Road, I imagined Darwin holding the ancient skull. He turns it with delicate hands and stares into the large, round eyes of the Gibraltar Neanderthal, rubbing his thumb against the thick, double-arched brow ridges. He marvels at the enormous size of the nasal cavity. Upon turning the skull to the side, he remarks to Falconer how the skull sweeps back and lacks the tall forehead of a modern human. Falconer reminds Darwin that just a year earlier, at the British Association for the Advancement of Science meeting, William King had presented evidence based on a partial skeleton from Feldhofer Cave in Neander valley, Germany, for an extinct population of Europeans he called *Homo neanderthalensis*. "One odd skull can be dismissed. But two? Two is a pattern," I imagine Darwin saying with a smile.

But probably none of that happened.

PREFACE. Gibraltar Neanderthal skull.
(© Chris Stringer/The Natural History Museum, London)

It is more likely that Darwin thanked Falconer for coming and apologized for his ill health, which had made him weak, unfocused, and at times depressed. Perhaps in this state, he could not focus on the Gibraltar skull without feeling faint and instead made a few cursory observations before carefully handing it back to Falconer.

Perhaps too it was difficult for Darwin to see the details of the Gibraltar skull that are so compelling to paleoanthropologists today. In 1864, the skull still had not been fully cleaned of its rocky matrix.[5] The details of the nasal cavity, for instance, were obscured by cemented sand. Or maybe he did study it carefully but recalled Thomas Henry Huxley's observations of the skull from the Feldhofer Cave. Huxley, whom Darwin trusted on matters related to human anatomy and evolution, wrote just a year earlier, in *Evidence as to Man's Place in Nature* (1863), that while the Feldhofer skull was unusual and ancient, it still fit within the range of variation of modern human skulls.

Darwin himself had been thinking quite a bit about variation, the source of variation, and the importance of variation in living populations. After he finished his manuscript on climbing plants, he was to devote himself entirely to the subject and eventually complete *The Variation of Animals and Plants under Domestication* in 1868.

Perhaps with variation on the brain, and Huxley's words echoing in his mind, he could encompass the Gibraltar cranium within the range of variation of modern human skulls. Darwin spent a lot more time studying barnacles, orchids, pigeons, and dogs than humans. Maybe he just hadn't seen enough human skulls to recognize the one from Gibraltar as different.

I imagine Falconer wished Darwin well, thanked Emma, and exited 4 Chester Place Road. Perhaps Darwin turned toward the window and watched his friend walk down Cumberland Place Road toward the park. Falconer's satchel was slung over his shoulder, and he cradled it, and its precious contents, with great care as he walked. Darwin looked past his friend, toward the botanical gardens, and wondered whether the plant tendrils had relaxed their grip on the trellis now that the wind had died down.

Science is done by scientists. Even the very best scientists in the world—in this case, Charles Darwin himself—err. Sometimes these errors are rooted in bias; sometimes they arise from insufficient data to answer the question being asked; sometimes they happen because an illness compromises focus when a friend visits with an ancient fossil skull one summer day in 1864.

Figuring out how the natural world works is not easy. Fossils do not come with labels. It took the discovery of many more fossils in the late nineteenth and early twentieth centuries for scientists to recognize the legitimacy of Neanderthals. Even then, it was not until DNA was extracted from these old bones that we began to truly understand the role of Neanderthals in human evolution. Today, we still have many unanswered questions about our extinct cousins.

The inner workings of the natural world do not magically reveal themselves; data never speak for themselves. Instead, interpretations of evidence are made by people who breathe meaning into empirically derived facts and figures. And scientific hypotheses are not always self-evident. They are generated in creative minds and tested by emotional, subjective humans behaving as objectively as they can. Science is thus a human endeavor.

Sometimes the humanity of science leads to great insights. But sometimes it leads to a scientist holding the evidence for a human past literally in his hands without recognizing it. That is why science cannot be done in isolation, by a single individual. It is a collective enterprise that unfolds over generations as we test and retest old ideas and develop new ones to make sense of our world. It stagnates when it is done by homogeneous scientists with similar backgrounds and experiences. It outright fails when it is practiced by inflexible individuals clinging desperately to tired ideas. Darwin knew this. "I had," he wrote, "during many years, followed a golden rule, namely that whenever published fact, a new observation of thought came across me, which was opposed to my general results, to make a memorandum of it without fail and at once; for I had found by experience that such facts and thoughts were far more apt to escape from the memory than favourable ones."[6]

While many of us scientists admire Darwin, we do not worship him. He was a brilliant scholar who not only generated new data but could see how his observations were connected under big, overarching ideas with both explanatory and predictive power. He navigated seamlessly between big picture, theory-level thinking and the small, intricate details. In studying orchids, earthworms, and barnacles, he could see both the forest and the trees. But Darwin had flaws, both as a scientist and as a human.

"I look with confidence to the future, to young and rising naturalists, who will be able to view both sides of the question with impartiality," he wrote.[7] The question he referred to, of course, was evolution or, as he called it, descent with modification. In 1859, in his most famous work, *On the Origin of Species*, Charles Darwin proposed a mechanism for biological evolution: natural selection. "Whilst this planet has gone cycling on,"[8] new generations of young and rising naturalists have indeed tested Darwin's ideas. Evolution by means of natural selection has been supported over and over again. Darwin was right.

Biological evolution is thus one of the most profound and influential scientific theories ever proposed. The implications of evolution are widespread and personal: all life on Earth is related and has changed over time. However, in *Origin*, Darwin made little mention of humans, noting only that "light will be thrown on the origin of man and his history."[9] But Darwin was indeed thinking about humans. He called human origins "the highest and most interesting problem for the naturalist." The title of this book is inspired by that line, which Darwin wrote in a letter to Alfred Russel Wallace on December 22, 1857.[10]

On February 24, 1871, Darwin published his thoughts on the human career in a two-volume compendium, *The Descent of Man, and Selection in Relation to Sex*. He wrote in the introduction, "It has often been asserted that man's origin can never be known: but ignorance more frequently begets confidence than does knowledge: it is those who know little, and not those who know much, who so positively assert that this or that problem will never be solved by science."[11] In other words, Darwin proposed that the origin of humans was knowable.

Yet, at that time, Darwin didn't know about DNA. The entirety of the human fossil record, which now numbers in the thousands of specimens, consisted of just a few Neanderthal bones. These bones were misidentified by most (including, as we proposed earlier, Darwin himself) as just unusual modern humans. Modern primatological studies of our great ape cousins were almost a century away. And scholars were even debating whether different races of humans had descended from different species of primate.

A lot has changed in 150 years.

Darwin was remarkably prophetic in some of his predictions—for example, that the earliest human fossils would be discovered on the African continent. But he was flat out wrong in other areas. That is how science works. Even the most elegant ideas can wither in the face of new evidence. Darwin did not present infallible statements to be revered in *Descent of Man*. He presented hypotheses to be tested. Some of these ideas have withstood 150 years of scrutiny. Some have not. Thus, *A Most Interesting Problem* is not so much a celebration of Darwin as it is a tribute to how science operates, how scientific ideas are tested, and the role of evidence in helping structure narratives of human origins.

On the 150th anniversary of the publication of *Descent of Man*, we present in these pages a view of where we are in our quest for understanding the origin, biological variation, behavior, and evolution of humans. Charles Darwin and evolution are inextricably linked. As Darwin's biographer Janet Browne, who pens the introduction to our book, has written, there is "much more to Darwin than his theory, and more to the theory than Darwin."[12] But Darwin serves as an appropriate benchmark for revisiting what we know and how we know it, and this anniversary provides an opportunity for self-reflection in our quest for understanding how humans evolved.

This book reviews, chapter by chapter, what Darwin wrote in the first edition of Part 1 and the last three chapters of Part 2 of *Descent*, comparing his words to what we now know 150 years later. The focus of this book is human evolution, and thus most of Part 2 of *Descent*, a detailed treatise on sexual selection, is covered only briefly in *A Most Interesting Problem*.

In 1871, the evidence for human ancestry could be found in comparative anatomical and embryological studies, updated for the reader in Chapter 1 by anatomist Alice Roberts. What we know today about the evolution of the human brain and the origins of morality and sociality are addressed in Chapters 2 and 3 by neuroanatomist Suzana Herculano-Houzel and biological anthropologist Brian Hare. The fossil evidence for our evolutionary history is explored in Chapter 4 by paleoanthropologist Yohannes Haile-Selassie. Bioarchaeologist Kristina Killgrove writes in Chapter 5 about Darwin's misguided conflation of biological and cultural evolution and the resulting rise of social Darwinism. In Chapter 6, paleoanthropologist John Hawks summarizes how molecular genetics has revealed our place in the primate family tree with resolution

Darwin could only dream of. In Chapter 7, anthropologist Agustín Fuentes modernizes our understanding of human races. As many have observed, *The Descent of Man, and Selection in Relation to Sex* is really two books in one, with Part 2 of *Descent* devoted primarily to Darwin's novel idea of sexual selection. Current understandings of sexual selection are presented in Chapter 8 by biologist Michael Ryan. Darwin's attempt to apply his ideas of sexual selection to humans are critiqued in Chapter 9 by anthropologist Holly Dunsworth. Science writer Ann Gibbons concludes the book with a summary of 150 years of scholarship in our discipline.

In these pages, topics Darwin first introduced 150 years ago are updated and expanded by ten scientists and science communicators currently studying these very questions and/or effectively and enthusiastically communicating the findings through public outreach and engagement. It was a deliberate choice to have contributors strong in science communication write the chapters in this book. Darwin himself wrote books for public consumption; it was fitting that we did the same.

And so, as you read each of the chapters, I invite you to imagine each of the authors taking a stroll with Charles Darwin around Down House's Sandwalk, the "thinking path" Darwin would walk each day as he struggled with and developed his scientific ideas. At the entrance to the path, Darwin was known to stack stones on top of one another and gently knock one over with his walking stick each time he passed by. A short walk needed but a few stones, a longer walk taken to contemplate bigger problems would require more. For this walk with Darwin, I welcome you to stack eleven stones—one for the introduction, and ten for the remaining chapters—and join us on this adventure.

Along the way, imagine Darwin learning about DNA and about how this elegant molecule not only solves his blending inheritance problem but reconstructs the family tree of life into nested hierarchies predicted by descent with modification. Imagine Darwin learning about *Australopithecus* and how we now know that upright walking and canine tooth reduction long preceded brain enlargement in the human lineage. On our walk, Darwin would be confronted with the compelling evidence that humans cannot be categorized into tidy boxes called races. The authors would enthusiastically tell Darwin about Hox genes, *Ardipithe-cus*, genetic drift, vitamin D, folic acid, and the evolution of skin coloration.

But our authors would not shy away from confronting him with data demonstrating how incorrect he was when he wrote that men were more intelligent than women. He would learn how his words were used to justify the eugenics movement of the early twentieth century. At some point in our walk, he would hear that in late August of 1864, he held in his hands the skull of a ~50,000-year-old extinct human called a Neanderthal. Perhaps he would stop walking, glance up in amazement, and shake his head with a smile. Given how intellectually curious Darwin was about the natural world, I suspect he would be thrilled at the discoveries of the past 150 years. I can only hope that he would also be deeply troubled by the use and misuse of his words to justify fallacious ideas of white male superiority that continue to this day.

Despite his ill health and frailty, I wonder whether, after eleven trips around the thinking path, Darwin himself would stack more stones at the Sandwalk entrance and continue the conversation long into the night. My hope is that after finishing this book, you too will continue on your own path, following the scientific exploration into the origins and evolution of our remarkable species.

Jeremy M. DeSilva
Hanover, New Hampshire
On the 150th anniversary of the publication of Charles Darwin's
The Descent of Man, and Selection in Relation to Sex.

CONTRIBUTORS

JANET BROWNE is the Aramont Professor of the History of Science at Harvard University. She has written a two-volume biography of Charles Darwin—*Darwin: Voyaging* and *Darwin: The Power of Place*. In 2013 she wrote the introduction to a republishing of Darwin's *Descent of Man*.

JEREMY M. DESILVA is an associate professor of anthropology at Dartmouth College. He studies the origins and evolution of upright walking in the human lineage. He worked at the Boston Museum of Science from 1998 to 2003 and continues to be passionate about science literacy. His book *First Steps: How Upright Walking Made Us Human* will be published 2021.

HOLLY DUNSWORTH is a professor of anthropology at the University of Rhode Island. Dunsworth challenges the traditional (often male-biased and Eurocentric) narratives of human evolution with exquisite clarity. She has contributed to NPR's *This I Believe* series and her posts to science blogs *The Mermaid's Tale* and *Origins* are must reads. She is writing a book tentatively titled *I Am Evolution*.

AGUSTÍN FUENTES is a professor of anthropology at Princeton University. He is the author of *Why We Believe: Evolution and the Human Way of Being* (2019), *The Creative Spark: How Imagination Made Humans Exceptional* (2017), and *Race, Monogamy, and Other Lies They Told You: Busting Myths about Human Nature* (2012). In 2018, he delivered the prestigious Gifford Lectures at the University of Edinburgh.

ANN GIBBONS is a contributing correspondent for *Science* magazine and the author of *The First Human: The Race to Discover Our Earliest Ancestors*. She has taught science writing at Carnegie Mellon University and written about human evolution for *National Geographic, Slate, Smithsonian* magazine, and other publications. She recently was awarded the 2019 American Geophysical Union's David Perlman Award for Excellence in Science Journalism.

YOHANNES HAILE-SELASSIE is the Curator and Head of Physical Anthropology at the Cleveland Museum of Natural History. He has made some of the most significant early human fossil discoveries in the history of paleoanthropology. In 2013, he helped design the stunning and innovative new Human Origins Gallery at the Cleveland Museum of Natural History.

BRIAN HARE is a professor of evolutionary anthropology at Duke University. His interest in the evolution of social behavior has inspired research on humans' closest ape relatives and humans' best friend, the dog. He has coauthored two books with Vanessa Woods: *The Genius of Dogs: How Dogs Are Smarter Than You Think* (a 2012 *New York Times* best seller) and *Survival of the Friendliest* (2020).

JOHN HAWKS is a professor of anthropology at the University of Wisconsin. His *John Hawks Weblog* (johnhawks.net) has been the go-to spot for paleoanthropology and human genetics news and analysis for the past decade. He codirected the first paleoanthropological fossil discovery to be live-tweeted and live-streamed, from the Rising Star Cave in South Africa. In 2017, he coauthored a book on this expedition, *Almost Human: The Astonishing Tale of* Homo naledi.

SUZANA HERCULANO-HOUZEL is an associate professor in the departments of Psychology and Biological Sciences at Vanderbilt University and the associate director for communications of the Vanderbilt Brain Institute. Her TED talk about the human brain has been viewed more than 3 million times and was featured on NPR's *TED Radio Hour* in 2015. In 2016, she published *The Human Advantage: How Our Brains Became Remarkable*.

KRISTINA KILLGROVE is a bioarchaeologist at the University of North Carolina, Chapel Hill, specializing in ancient Rome. She writes about bioarchaeology for *Forbes* magazine, where her contributions are read by millions of people every year. Her first book, *These Old Roman Bones*, is in progress.

ALICE ROBERTS is a biological anthropologist and clinical anatomist. She is the Professor of Public Engagement of Science at the University of Birmingham, United Kingdom. Her BBC documentaries *The Incredible Human Journey* and *Origins of Us* have been watched by millions. She has written seven books on human anatomy and evolution, including *The Incredible Unlikeliness of Being*, published in 2015.

MICHAEL J. RYAN is the Clark Hubbs Regents Professor in Zoology at the University of Texas at Austin and a senior research associate at the Smithsonian Tropical Research Institute in Panama. He has given more than 150 invited lectures on his research, and his book *A Taste for the Beautiful: The Evolution of Attraction* was published by Princeton University Press in 2018.

A MOST INTERESTING PROBLEM

Introduction

Janet Browne

IN THE DESCENT OF MAN, Charles Darwin dealt with what he called "the highest & most interesting problem for the naturalist." This volume of essays shows how true these words still remain in the twenty-first century. Published in 1871, *The Descent of Man, and Selection in Relation to Sex** was a comprehensive statement of Darwin's theory of evolution as it applied to human beings and a far-reaching account of the biological phenomenon that he termed sexual selection; in it Darwin described what he knew about human ancestral origins, the physical characteristics of different peoples, the emergence of language and the moral sense, the relations between the sexes in animals and in humans, and a host of similar topics that blurred the boundaries between ourselves and the animal world. His aim was to demonstrate that human beings had gradually evolved from animals and that the differences were only of degree, not kind. His conclusions were bold: "We must acknowledge, as it seems to

---

* The first edition, numbering 2,500 copies, was published on February 24, 1871. There are two issues of this edition. The first issue can be distinguished by the inclusion of a note about errata. The printer evidently corrected these errata in the second issue, of 2,000 copies, released in March 1871. The book cost one pound four shillings in a standard green binding. Darwin's own copy, however, is dated 1870 and was evidently in his hands direct from the printer in December 1870. Richard Freeman, *The Works of Charles Darwin: An Annotated Bibliographical Handlist*, 2nd ed. (Folkestone, England: Dawson, 1977). *Descent* was the first book from which Darwin gained a financial profit. Publisher John Murray sent Darwin a check for £1,470.

me, that man, with all his noble qualities, with sympathy which feels for the most debased, with benevolence which extends not only to other men but to the humblest living creature, with his god-like intellect which has penetrated into the movements and constitution of the solar system—with all these exalted powers—Man still bears in his bodily frame the indelible stamp of his lowly origin."[1]

It had been a long process that brought Darwin to this point. Twelve years earlier, in *On the Origin of Species by Means of Natural Selection, or the Preservation of Favoured Races in the Struggle for Life*, he had cautiously written that if his views were accepted, "light would be thrown on the origins of man and his history."[2] Since that time, controversy about the possibility of a natural origin for all living beings had swept the globe wherever science was seen as a modernizing force. Was Darwin seriously suggesting that the natural world had not been created by any form of divine being? Were the Judaic and Christian stories of Adam and Eve to be jettisoned? Was there convincing evidence for natural selection? And how could such a mechanical process, ultimately based on probabilities, produce the extraordinary adaptations of animals and plants, let alone the moral sense, language, or civilizations of human beings?

These were some of the issues that are now often referred to as the Darwinian controversies over science and religion. In the nineteenth century, frontline issues largely drew on strong antipathy to the idea of human descent from animals. Christian critics objected to losing the central doctrines of the Bible, even though the Genesis story was already mostly seen as an allegory, and believers in other faiths similarly recoiled from giving up the special place of human beings in nature. Yet, the debate spread further than religious controversy. Secular thinkers criticized the theory on practical grounds: Where was the evidence, and how could it work without some teleological organizing principle at its heart? Debates flared over the prospect of science providing answers to questions that had traditionally been the preserve of theologians and philosophers. To many, Darwin's ideas heralded a new form of scientific naturalism that could transform the status of science in the Victorian world.[3] Radicals saw in it the possibility of atheism and the overthrow of contemporary political hierarchies.[4] Throughout, the puzzle of human diversity and what was then called "racial science" pervaded imperial rhetoric and drew on evolutionary theory for support.

The Descent of Man, and Selection in Relation to Sex was Darwin's carefully considered response to such questions. In its way, this book on humankind was just as memorable as *Origin of Species* and can perhaps be regarded as *Origin*'s missing half. The word "evolution," first used in its modern sense, occurs on page two of the first volume of *Descent*, in the introduction, where Darwin discussed the likelihood of natural scientists accepting the idea of natural selection.[5] Darwin also used the term "survival of the fittest," which he had adopted from Herbert Spencer in 1868.[6] *Descent of Man* was written in the same personal style as *Origin of Species*, with the same courteous modesty, the same clarity, the same inexhaustible piles of evidence, and the same explicit rationalism. Its intellectual breadth was astonishing. And even though the format now seems archaic, the style of reasoning overly anecdotal, and the social views regrettably typical of a nineteenth-century British gentleman, Darwin's central arguments retain, even today, their power to explain aspects of the natural world, as is amply shown in this volume.

Darwin's book was issued in two volumes and contained two parts, as indicated in the title. In Part 1, Darwin gave a systematic account of the connections between humans and animals. He covered comparative anatomy and, at much greater length, the human mental faculties—language, reasoning ability, morality, consciousness, the religious sense, memory, and imagination. Everything that characterized the mind of human beings, Darwin posited, had emerged from animal ancestors, stepwise, by entirely natural processes. In Part 2, Darwin presented his important new concept of sexual selection. He explained how this was different from natural selection and how it worked as a complementary force in evolutionary change. Much of this section was dedicated to establishing what he meant by sexual selection and necessarily included lengthy discussion of the process as discerned in animals. At the end of Part 2, Darwin proposed that sexual selection was instrumental in explaining the origin of what he called human "races" and cultural progress. Here there are fascinating glimpses into Darwin's understanding of the biological basis of Victorian racial hierarchies, gender relations, and the structure of contemporary civilization.

Born into an Industrial, Colonial Age

Descent of Man shows Darwin at his most Victorian. His life (1809–1882) spanned much of the nineteenth century, and his science reflected the industrial and political transformations for which Britain was then famous.[7] From his childhood, he absorbed the prevailing ideology of industrial and colonial progress. He was born in the British industrial Midlands, in the town of Shrewsbury, to a prosperous medical family. One of his grandfathers was Josiah Wedgwood, the chinaware manufacturer, who was a leading figure in the British industrial revolution and the antislavery movements of the day. Wedgwood transformed the consumer market with his factory-produced chinaware but also participated in developing new manufacturing operations, applying ideas such as the division of labor, and opened up the British transport infrastructure by investing in canals and roads. Much of the financial and social capital on which the family's prosperity rested derived from Josiah Wedgwood's commercial success. Darwin's other grandfather was the prominent physician, liberal thinker, and literary figure Dr. Erasmus Darwin. Erasmus Darwin was a member of the small circle of "improvers," medical men and politically progressive intellectuals, who called themselves the Lunar Society.[8] The Darwin-Wedgwood circle was passionate in its support for abolitionism, and the young Charles Darwin adopted this frame of mind. The family's intellectual pursuits, along with its professional social standing, religious skepticism, high levels of education, commercial acumen, and liberal political views ensured that Charles Darwin always had a place in intellectual British society as well as the prospect of a comfortable financial inheritance, both of which were material factors in his later achievements.

More generally, during Darwin's lifetime, great currents of change were also making their presence felt. In the 1830s, the British nation came as close to political revolution as it ever did, owing to conflict: between landlords and manufacturers, workers against masters, province versus metropolis, the hungry and mutinous threatening the commercial-minded and individualistic middle classes. Prime minister Benjamin Disraeli's imagery of two nations, rich and poor, was not over-fanciful. The century had opened with warfare. At the end of the century, Britain was again at war, this time in South Africa. By then, imperial expansion and the second industrial revolution—marked by the coming of the railways, significant urbanization, the rise of the middle classes, increasing prosperity, and widespread dissemination of printed texts—was well under way. From the 1850s, a new and varied economy soaked up excess capital, leading to a diversification in the labor force. And in religious terms, although the Protestant (Anglican) faith provided the structure within which most British people operated, the grip of the church was loosening. Dissenting and nonconformist Protestant groups claimed the right to worship in their own manner, to educate the young, to be represented in Parliament, and to take public position and have their views heard. The foundation of a nondenominational University College in London in 1826 marked the opening of higher education to every citizen regardless of creed.

In science, matters were similarly expanding, diversifying, and refocusing. One by one, Victorian thinkers aimed to investigate the world around them without recourse to the Bible's word or the church's doctrinal authority. Religious doubts, secular inclinations, and dissatisfaction with conventional religious doctrines, especially the prevailing system of natural theology, were launched among British intellectuals long before Darwin came on the scene. There was as well rising engagement with science among many different groups of the British public.[9]

By the time Darwin published *Origin of Species*, the nation was witnessing industrial diversification, commercial and professional specialization, religious tension, intense colonial activity, and among the middle classes much talk of national "improvement" and "progress." The self-congratulatory sense of the era was captured by the Great Exhibition of the Works of Industry of All Nations, held in 1851 in central London, in the giant glass exhibition hall dubbed the Crystal Palace.

So why did Darwin deliberately choose to keep humankind out of *Origin of Species*? No doubt he was cautious about stirring up too much controversy. As he explained to Alfred Russel Wallacein in 1857, some years before publication, "You ask whether I shall discuss 'man'; I think I shall avoid whole subject, as so surrounded with prejudices, though I fully admit that it is the highest & most interesting problem for the naturalist."[10] But perhaps also because there was widespread middle-class unease about any social, political, or intellectual activities that threatened the status quo. Among these threats were notions of self-generated evolution or, as it was then called, transmutation—that is, change and progress without any divine creation or oversight. To adopt transmutation, as was seen with Robert Chambers's *Vestiges of the Natural History of Creation*, published anonymously in 1844,[11] or to promote points of view that advocated self-determination among human beings, such as the doctrine of phrenology,[12] was at that time to brand oneself as a dangerous political radical who might favor materialism and political upheaval. Radical thinkers might find biological support in transmutation for rejecting the hierarchical social structure of the United Kingdom and thus destabilize the state.

Darwin's Early Views on Humankind

Descent of Man can be regarded as the completion of an intellectual project begun during Darwin's travels on the *Beagle* nearly forty years beforehand.

Today the fame of the *Beagle* voyage sometimes makes it hard to remember that its purpose was not to take Darwin around the world but to carry out British Admiralty instructions. The ship had been commissioned to extend an earlier hydrographic survey of South American waters that had taken place from 1825 to 1830. The area was significant to the British government for commercial, national, and naval reasons, buttressed by the Admiralty's preoccupation with providing accurate sea charts and safe harbors for its fleet in the world's oceans. Robert FitzRoy, commissioned as the captain of the expedition, invited Darwin to join the voyage as an accompanying man of science, a "gentleman naturalist," who would collect natural history specimens, make observations and excursions as possible, and very generally keep the captain

FIGURE I.1. Fuegians encountered during the *Beagle* voyage, in 1835. Over thirty years later, Darwin wrote, in *Descent of Man*: "The astonishment which I felt on first seeing a party of Fuegians on a wild and broken shore will never be forgotten by me, for the reflection at once rushed to my mind—such were our ancestors." (Engraving by T. Landseer after Conrad Martens, frontispiece of Robert FitzRoy's *Narrative of the Surveying Voyages of HMS* Adventure *and* Beagle, vol. 2, 1839)

company. This curious relationship was unusual in the history of exploration. It also meant that Darwin's voyage was often a voyage on land. He had no duties on board. He could arrange whenever possible to be dropped off and picked up at various points, and he made several long inland expeditions in South America with hired guides, including a daring trek across the Andes.

Many aspects of the five-year voyage contributed to Darwin's emerging wish to understand the interconnections of living beings, although his visit to the Galápagos archipelago is the focus of most accounts. Retrospectively, however, it can be seen that his encounters with indigenous peoples were also important elements in unsettling his ideas about the stability of the natural world. The most significant of these was his experience with the inhabitants of Tierra del Fuego (Figure I.1). On board the *Beagle* were three individuals who had been taken from Tierra del Fuego to England by Captain FitzRoy on the previous *Beagle* voyage and were now being repatriated to serve in a projected Protestant mission station to be set up in the far south. FitzRoy had educated these three, and they had become relatively anglicized during their enforced stay in London. Darwin was fascinated by the returning Fuegians, especially O'rundel'lico (or Jemmy Button, as FitzRoy renamed him). In his diary, Darwin recounted his naïve amazement that, after so few years in English company, the three on board the ship were now almost another "species of man" from their literal relatives. This encounter encouraged him to think that human beings could be examined in scientific terms, as part of natural history, in the same way as other species. "I could not have believed," he wrote in his *Journal of Researches* after the voyage ended, "how wide was the difference, between savage and civilized man. It is greater than between a wild and domesticated animal, in as much as in man there is a greater power of improvement."[13]

For two decades after he returned from the *Beagle* voyage, Darwin kept notes about human evolution and pondered how best to develop and present his views.[14] Nevertheless, he chose to foreground other themes in the research program he undertook in preparation for publishing and deliberately kept humankind out of *Origin of Species*. He must have felt justified in some way when the storm of controversy erupted after publication of *Origin* over the possible apish origins of humankind. Such opinions were dramatized in England in 1860, in a public confrontation between the youthful naturalist Thomas Henry Huxley and conservative theologian Samuel Wilberforce, the bishop of Oxford. The confrontation (which was apparently unplanned) occurred at the annual meeting of the British Association for the Advancement of Science and is remembered today for a clever verbal exchange in which Wilberforce supposedly asked Huxley whether it was through his grandfather or his grandmother that he claimed his descent from an ape. Huxley is thought to have replied that he would not be ashamed to have a monkey for his ancestor but would not wish to be connected with a man who used his great intellectual gifts to obscure the truth. No verbatim account of the discussion exists, and there is considerable uncertainty regarding what Huxley and Wilberforce actually said. But the moment quickly came to symbolize the divergent positions that were being taken on human origins, with the Christian church, as represented by the bishop, standing firm on the divine creation of humankind, and science, as represented by Huxley, offering an entirely naturalistic alternative.[15]

So, Darwin bided his time. After *Origin of Species* was published, he threw himself into research projects that illustrated the concept of natural selection in ways that did not relate to human ancestry. He published a careful study of the fertilization of orchids by insects in 1862 that explored coadaptation in depth and an extensive analysis of the variation of animals and plants under domestication in 1868. It is important for us today to recognize Darwin as a superb experimental naturalist. But perhaps these projects also allowed him to evade harder questions, although he always considered such projects as vital supporting evidence for his theory.

Writing *Descent of Man*

Indeed, Darwin might never have published his ideas on humankind if it had not been for changing circumstances brought about by the controversies surrounding *Origin of Species* and the writings of some of his contemporaries on that very issue.

Soon after the publication of *Origin of Species*, three of Darwin's clos-est scientific friends produced important studies that developed differ-ent aspects of evolution as it related to humankind. In 1863, Charles Lyell published the *Antiquity of Man*. In this book, Lyell described the long course of human geological history. Lyell did not have much infor-mation to give on actual fossilized humans or prehumans—there were only a few broken parts of crania in collections at that time, and their identification was contested (we now know that they were among the earliest discovered remains of Neanderthals). He focused instead on prehistoric humankind—cavemen and -women. Until then, the paucity of early human artifacts such as worked flints and tools had suggested that humankind was very recent in geological terms, a view that ac-corded well with the notion that humans had appeared on Earth only when the Bible story started, some 4,000 years ago. Even those who believed in a longer age for the habitable Earth, including those few who believed in non-divine origins for humanity, were sure that human his-tory was relatively short and could be measured in mere thousands of years, not whole geological epochs. The common assumption was that humans appeared only when the planet arrived at its modern state, which was presumed to be after the glacial period—or, for those who believed in the biblical flood, at the point when the floodwaters receded. Lyell pushed the origin of humankind further back, beyond this watery dividing line, into the geological deep past. It was the first significant book after Darwin's *Origin of Species* to shake the contemporary view of humanity.

Then came Thomas Henry Huxley's book, *Evidence as to Man's Place in Nature*, which was published a few weeks after Lyell's. The text showed Huxley at his most argumentative. He used this small volume to con-tinue a scientific dispute with the great anatomist Richard Owen on the anatomical similarities between apes and humans. Partly, too, he used the opportunity to pioneer secular natural history and consolidate his rising position as the main public protagonist for Darwin.[16] Even though Huxley did not fully adopt Darwin's ideas, he defended Dar-win's right to propose entirely naturalistic explanations for the living world. In this short, polemical book, Huxley demonstrated how humankind must, on all biological grounds, be classified with the apes. The frontispiece (drawn by Benjamin Waterhouse Hawkins) showed five primate skeletons standing in line, each figure leaning slightly forward, ready to evolve into the next. The scale was cleverly adjusted to make the point. From gibbon to orangutan, chimpanzee, gorilla, and human, the implication could not be clearer—humans were the result of a series of physical changes from the apish state. The point was understood by readers but not necessarily accepted. One reviewer observed dryly, "We are not yet obliged to be quite on all-fours with Professor Huxley."[17]

Soon after Huxley was Alfred Russel Wallace, who had formulated the principle of evolution by natural selection independently of Darwin. Wallace wrote two compelling articles on human evolution in the 1860s. In the first, saying what Darwin had stopped short of saying in *Origin*, he argued that natural selection was the primary force in changing apes into people. In the second article, published in the 1869 *Quarterly Review*, Wallace backtracked and declared that natural selection seemed to him insufficient to explain the origin of humankind's extraordinary mental capacities. He agreed with Darwin that natural selection pushed our apish ancestors to the threshold of humanity. But at that point, he thought, physical evolution stopped and something else took over—the power of mind. The human mind alone continued to advance, human societies emerged, and cultural imperatives took over. According to Wallace, not every society developed at the same rate, accounting for what he and his contemporaries considered to be visible differences in cultural status. Darwin was thoroughly taken aback. "I hope you have not murdered too completely your own and my child," he wrote to Wallace in surprise.[18] Darwin's view was that everything that could be considered characteristic of the human condition—language, morality, religious sense, maternal affection, civilization, appreciation of beauty— had emerged in gradual steps from animals. He could not agree with Wallace that some external force—Wallace believed it to be some spiritual power—had made us what we are.

Other publications on human origins were evident too. The creative evolutionism espoused by George Campbell, the Duke of Argyll, was gaining ground. Herbert Spencer's *Principles of Biology* (1864) and

FIGURE 1.2. Darwin's study in his home, Down House, in Kent, United Kingdom. Here Darwin wrote his most famous books, including *Descent of Man* and *Origin of Species*. (Photograph by Jeremy DeSilva)

Essays: Scientific, Political and Speculative (1868) integrated evolutionary concepts with political, social, and religious ideas that were already attractive to contemporaries. Ernst Haeckel was busy describing apish ancestry to a German-speaking readership.

The moment at last seemed ripe to Darwin for completing his research on humankind and making it public. He could call on the investigations of prominent anatomists and anthropologists who were favorably disposed toward a secular, biological view of humankind. He was able to consult scientific contemporaries such as Francis Galton, John Lubbock, and Edward B. Tylor, and reach out to knowledgeable colleagues like Haeckel, Pierre Paul Broca, Jean Louis Armand de Quatrefages, Édouard Claparède, and Carl Vogt. His immense network of correspondents could help in locating specialists to guide him through relatively unfamiliar areas, such as the likely beginnings of human language, and in gathering further information on a mass of topics from individuals across the globe. The study in his house in Kent was his center of operations—arranged to offer a private and active working space for his many different projects (Figure I.2). He asked his daughter Henrietta Darwin, age twenty-eight, to act as copy editor and proofreader, to correct his grammatical mistakes and help with clarity. Soon Darwin had gathered so much material that he felt obliged to put some of it aside for another book. This additional material concerned the expression of emotions in animals and humans and was published in 1872, one year after *Descent of Man*, under the title *The Expression of the Emotions in Man and Animals*. These two books represent Darwin's most important statements on the evolution of humankind.

There was a lot for Darwin to keep in mind, a lot to reformulate and squeeze into shape. "I shall be well abused," he remarked to his close friend Joseph Hooker just before publication, in February 1871.

Publishing *Descent of Man*

John Murray, the publisher of most of Darwin's previous books, flinched at the subject matter of the scientist's latest. Despite his familiarity with Darwin's unorthodox topics and his determination not to let them stand in the way of a successful business relationship, this book on human ancestry rattled his confidence rather more than *Origin of Species* had done. He asked his publishing colleague, the Reverend Whitwell Elwin, for a professional opinion on Darwin's manuscript. Elwin was the former editor of Murray's *Quarterly Review* and often served as a useful barometer of public opinion for the publisher. "The arguments in the sheets you have sent me appear to me to be little better than drivel," Elwin groused.[19]

Murray bravely went ahead. He printed 2,500 copies of *Descent of Man*, publishing the book in early 1871. Three further print runs were issued during the same year, bringing the number of copies available to readers up to 8,000. Darwin made small changes in the texts of each reprint. For bibliophiles, there are some interesting variants. Darwin's own copy, for example, was ready by December 1870 and has that date printed on its title page. Murray published a second edition in 1874 with corrections and emendations. By 1877 Murray's firm recorded that it had issued a total of 11,000 copies. The American publishing house of D. Appleton and Company simultaneously published *Descent of Man* in New York in 1871 and continued to match the English editions pretty closely. In Europe, the Franco-Prussian War would seemingly have obliterated any prospect of overseas editions and foreign translations. Yet—astonishingly, in view of the political situation, especially during the Siege of Paris and the dreadful events around the Commune of Paris—Darwin's book was translated into Dutch, French, German, Russian, and Italian in 1871, and into Swedish, Polish, and Danish shortly thereafter, a testimony to the fortitude of Darwin's European colleagues and general interest in evolutionary affairs.

Darwin began *Descent of Man* by relating the many incontrovertible anatomical features common to both animals and humankind. Part of his point was to establish that human beings are just as variable in their physical constitution as animals—a continuation of his comparative argument from *Origin of Species*. Then he turned to the mental powers, stating decisively, "There is no fundamental difference between man and the higher mammals in their mental faculties."[20] He presented anecdotal observations of animal behavior in substantiation of this claim, with examples ranging from horses that knew the way home to ants that defended their property, chimpanzees that used twigs as implements, bowerbirds that admired the beauty of their nests, and household cats and dogs that apparently dreamed of chasing rabbits in their sleep. The domestic nature of Darwin's observations in this area, the large doses of willing anthropomorphism, his evident delight in traditional country pursuits, and the glimpses he provided of the congenial home life of a Victorian gentleman inspired Frances Power Cobbe to deride these accounts as "fairy tales of science," in a review published in 1872.[21] These anecdotes probably went a long way toward softening his readers before he confronted them with the shock of apes in the family tree (Figure I.3).

To explain the emergence of the mind and language of humankind through variation and natural selection was altogether more problematic. Darwin launched straight into an examination of the power of human speech: this was obviously critical for him, since language was integral to all contemporary definitions of being human and was assumed to present an inseparable barrier between animals and humans. Darwin particularly wished to contest the widespread view that the ability to speak indicated God's special gift to humans. The great linguist and scholar Friedrich Max Müller had expressed the view that human language was a divine gift in the magazine *Nature* in 1870. Darwin believed that the ability to speak must have emerged quite differently, arising in a gradual fashion from the social vocalizations of apes and further developing in extremely early human societies through the imitation of natural sounds.[22]

Darwin was similarly daring when dealing with the evolution of religious belief. Drawing on the work of the cultural anthropologist Edward B. Tylor, he mapped out a comparative evolution of the religious sense, proposing that religious belief was ultimately nothing more than an urge to bestow a cause on otherwise inexplicable natural events. He proposed that human dreams occurring in early societies might have given rise to the idea of external gods, as Tylor suggested, or to animism, in which plants and animals seem as if they are imbued with spirits. Darwin suggested that these beliefs could easily grow into a conviction about the existence of one or more gods who directed human affairs. As

FIGURE I.3. "A Venerable Orang-Outang." Caricature of Charles Darwin issued after *Descent of Man* was published. (From *The Hornet*, March 22, 1871)

societies advanced in civilization, he said, ethical values would become attached to such ideas. "Strange superstitions and customs" would give way to the "improvement of reason, to science, and our accumulated knowledge." Human beings have a biological need to believe in something "other," he suggested. Audaciously, he compared religious devotion to the "love of a dog for its master."[23]

As for human morality, Darwin pointed out that the concept was only relative. Careful reading in canonical moral philosophy texts and long observational experience with household pets (and no doubt his children as well) told him that living beings had to learn the difference between "good" and "bad" behavior—the knowledge was not innate. Moreover, members of what he called "primitive" societies held a wide range of ideas about acceptable behavior, many of which he knew would horrify contemporary Victorians, such as cannibalism. If honeybees ever became as intelligent as humans, Darwin explained, unmarried females would think it a "sacred duty to kill their brothers, and mothers would strive to kill their fertile daughters; and no one would think of interfering."[24]

Darwin proposed this more for effect than logical necessity, because he went on to argue that higher human values emerged and spread only as human civilization progressed, meaning that duty, self-sacrifice, virtue, altruism, and humanitarianism were acquired fairly late in human history and not equally by all tribes or groups. "How little the old Romans knew of [sympathy] is shewn by their abhorrent gladiatorial exhibitions. The very idea of humanity, as far as I could observe, was new to most of the Gauchos of the Pampas."[25] It is clear that Darwin thought there had been a progressive advance of moral sentiment from the ancient "barbaric" societies described in Victorian history books, such as those of ancient Greece or Rome, to the civilized world of nineteenth-century England that he inhabited. In this manner, he kept the English middling classes to the front of his readers' minds as representative of all that was best. The higher moral values were, for him, self-evidently the values of his own class and nation.

Even the sense of duty was for Darwin biologically based in the social instincts. "The highest stage in moral culture at which we can arrive," he wrote, "is when we recognise that we ought to control our thoughts."[26] To be sure, Darwin praised the intrinsic nobility of this moral feeling, quoting Immanuel Kant. "Duty! Wondrous thought, that workest neither by fond insinuation, flattery, nor by any threat . . . whence thy original?"[27] Yet as Darwin described it, a female monkey who voluntarily sacrificed herself for her offspring would not only ensure her children's survival but also supply the next generation with the hereditary material (Darwin had no notion of modern genetics or kin selection) that favored such action again. Personally, he declared, he would rather be descended from a heroic little monkey that sacrificed her life in this manner than from a savage "who delights to torture his enemies, offers up bloody sacrifices, practises infanticide without remorse, treats his wives like slaves, knows no decency, and is haunted by the grossest superstitions."[28]

In Part 1, Darwin also discussed fossil intermediaries between ape and human and mapped out (in words) a provisional family tree, in which he took information mostly from fellow evolutionists such as Ernst Haeckel and Thomas Henry Huxley. In truth, Darwin found it difficult to apply an actual evolutionary tree to humans. Briefly, he tracked humans back as far as the Old World monkeys, saying that the human species must have diverged from the original monkey stock considerably earlier than did the anthropoid apes, probably at a point close to now-extinct forms of Lemuridae. He recognized the great apes as humanity's nearest relatives. Darwin knew very little about fossil primates and could name only *Dryopithecus*, the largest fossil ape identified in the deposits of Europe at that time (for the second edition of *Descent of Man*, Darwin asked Huxley to fill this gap with an up-to-date essay about fossil finds). He could only guess at possible reasons for human ancestral forms to have abandoned the trees, lost their hairy covering, and become bipedal.

> The early progenitors of Man were no doubt once covered with hair, both sexes having beards; their ears were pointed and capable of movement; and their bodies were provided with a tail, having the proper muscles. . . . The foot, judging from the condition of the great

toe in the foetus, was then prehensile; and our progenitors, no doubt, were arboreal in their habits, frequenting some warm, forest-clad spot. The males were provided with great canine teeth, which served them as formidable weapons.[29]

Sexual Selection and Society

An important part of *Descent of Man* was Darwin's account of human racial diversification. He believed that sexual selection held the answers. "I do not intend to assert that sexual selection will account for all the differences between the races," he wrote in *Descent*.[30] Nonetheless, he told Wallace in a letter in 1867 that he felt certain that it was "the main agent in forming the races of man."[31] As early as 1864 he had in fact explained to Wallace that sexual selection could be "the most powerful means of changing the races of man that I know."[32]

In defining sexual selection, Darwin postulated that all animals, including human beings, possess many trifling features that are developed and remain in a population solely because they contribute to reproductive success. These features were heritable (as Darwin understood it) but carried no direct adaptive or survival value. The classic example is the male peacock that develops large tail feathers to enhance its chances in the mating game, even though the same feathers actively impede its ability to fly away from predators. The female peahen, argued Darwin, is attracted to large showy feathers and, if she can, will choose the most adorned mate and thereby pass his characteristics on to the next generation. It was a system, he stressed, that depended on individual choice rather than survival value. Darwin devoted nearly one-third of *Descent of Man* to establishing the existence of this sexual selection in birds, mammals, and insects. In animals, he argued, the choice of mate was determined by the female: the female peahen did the choosing. When he came to humans, he reversed that proposition and insisted that men did the choosing.[33]

Darwin used sexual selection to explain the divergence of early humans into the racial groups that Victorian physical anthropologists described. Skin colors were for him a good example. Early men, he suggested, would choose their mates according to localized ideas of beauty. As the men in any group continued to express their preference for one or another ideal of beauty in women, so the external character-istics of the population would shift. "The strongest and most vigorous men . . . would generally have been able to select the more attractive women . . . who would rear on an average a greater number of children."[34] Each society would have dissimilar ideas about what con-stituted attractiveness, and so the physical features of various groups would gradually diverge through sexual selection alone.

Darwin argued that sexual selection was not confined to physical attributes such as hair or skin color. According to him, sexual selection among humans would also affect mental traits such as intelligence, ma-ternal love, bravery, altruism, obedience, and the "ingenuity" of any given population; that is, heterosexual human pairing choices would go to work on the basic animal instincts and push them in particular directions.

These views were utterly embedded in Darwin's personal social cir-cumstances. While he made a good attempt to be culturally relativistic, he still drew on the conventional ideas of his era and social position about human pairing behavior, choice, and gender. For example, he be-lieved that sexual selection had fostered built-in male superiority across the world. In early human societies, he argued, the necessities of sur-vival had resulted in men becoming physically stronger than women and in their intelligence and mental faculties improving beyond those of women. In civilized regimes it was evident to him that men, because of their well-developed intellectual and entrepreneurial capacities, ruled the social order.

In this way Darwin made human society an extension of biology and saw in every human group a "natural" basis for primacy of the male. After *Descent of Man*'s publication, early feminists and suffragettes bit-terly attacked this doctrine, feeling that women were being "naturalized" by biology into a secondary, submissive role.[35] Indeed, many medical men asserted that women's brains were smaller than those of men, and they were eager enough to adopt Darwin's suggestion that women were altogether less evolutionarily developed and that the "natural" function of women was to reproduce, not to think. For several decades, Anglo-American men in the medical profession thought that the female body was especially prone to medical disorders if the reproductive functions were denied. Something of this belief can be traced right through to the 1950s and beyond.

In *Descent of Man*, Darwin also made concrete his thoughts on human cultural progress and civilization. The notion of a hierarchy of races informed his discussion and took added weight from being published at a time when the ideology of extending one nation's rule over other nations or peoples was unquestioned. Darwin stated that natural selection and sexual selection combined with cultural shifts in learned behavior to account for the differences that he saw between populations. The racial hierarchy, as Darwin called it, ran from the most primitive tribes of mankind to the most civilized and had emerged over the course of eons through competition, selection, and conquest. Those tribes with little or no culture (as determined by Europeans) were, he thought, likely to be overrun by bolder or more sophisticated populations. "All that we know about savages, or may infer from their traditions and from old monuments," he wrote, "shew that from the remotest times successful tribes have supplanted other tribes."[36] Darwin was certain that many of the currently existing peoples he called primitive would in time similarly be overrun and perhaps destroyed by more advanced races, such as Europeans; he had in mind particularly Tasmanian, Australian, and New Zealand aboriginal peoples. This to him was the playing out of the great law of "the preservation of favoured races in the struggle for life," as expressed in the subtitle of his earlier book *On the Origin of Species*. Such an emphasis on the natural qualities underpinning social cultural development explicitly cast the notion of race into biologically determinist terms, reinforcing then contemporary ideas of a racial hierarchy.

Partly because of Darwin's endorsement and partly because of the influential writings of others, these views intensified during the high imperialism of the early twentieth century. Herbert Spencer's doctrine of "survival of the fittest," as used by Darwin, Wallace, Spencer, and others, in *Descent of Man* and elsewhere, became a popular phrase in the development of social Darwinism. Embedded in powerful class, racial, and gender distinctions, social Darwinism used the prevailing ideas of competition and conquest to justify social and economic policies in which prosperity and success were the exclusive aim.[37] "Survival of the fittest" was a phrase well suited to encourage hard-nosed economic expansion, rapid adaptation to circumstance, and colonization. Karl Pearson, a committed Darwinian biologist, expressed it starkly in Britain in 1900: no one, he said, should regret that "a capable and stalwart race of white men should replace a dark-skinned tribe which can neither utilise its land for the full benefit of mankind, nor contribute its quota to the common stock of human knowledge."[38]

Several of Darwin's remarks in *Descent of Man* captured anxieties that were soon to be made manifest in the eugenics movement. Darwin feared that what he called the "better" members of society were in danger of being numerically swamped by the "unfit." In this latter category Darwin included men and women of the streets, the ill, indigents, alcoholics, and those with physical disabilities or mental disturbances. He pointed out that medical aid and charity given to the sick and the poor ran against the fundamental principle of natural selection. Evidently torn between his social conscience and what he understood about evolutionary biology, he went on to declare that it was a characteristic of a truly civilized country to aid the sick and help the weak.

In these passages Darwin anticipated some of the problems that his cousin Francis Galton would try to alleviate through the eugenics movement. Galton was an enthusiastic convert to Darwin's theories and had little hesitation in applying the concept of selection to human populations. He aimed to improve human society though the principles of natural selection: in essence, by reducing the rate of reproduction among those he categorized as the poorer, unfit, profligate elements of society and promoting higher rates of reproduction among the middle classes. Galton hoped that the men he called highly gifted—the more successful men—should have children and pass their attributes on to the next generation. Galton did not promote policies of incarceration or sterilization ultimately adopted by the United States, nor did he conceive of the possibility of the whole-scale extermination of "undesirable" groups as played out during World War II. But he was a prominent advocate of taking human development into our own hands and the necessity of improving the human race. Darwin referred to Galton's point of view in *Descent*.

While Darwin's *Descent of Man* can hardly account for all the racial stereotyping, nationalist fervor, and prejudice expressed in years to come, there can be no denying the impact of his work in providing a biological backing for notions of racial superiority, reproductive constraints, gendered typologies, and class distinctions.

Legacy

Darwin's *Descent of Man* could nowadays be considered something of a period piece in the style of argument, the use of evidence, and the conclusions put forward. Yet, as this volume of essays shows, it opened one of the first genuinely public debates about human origins to stretch across general society. The critiques, scientific responses, and thoughtful debates originally generated were evocative of the social diversity of the nineteenth century and remind us that the introduction of new and culturally difficult ideas is rarely straightforward. Moreover, Darwin's book encouraged important long-term further investigation, both in the lab and in the field, for many different audiences and in many different languages. This continuing work is a remarkable tribute to the lasting power of Darwin's vision and the ideas themselves.

1

The Fetus, the Fish Heart, and the Fruit Fly

Alice Roberts

The sole object of this work is to consider, firstly, whether man, like every other species, is descended from some pre-existing form.

CHARLES DARWIN, *THE DESCENT OF MAN*[1]

IN "THE EVIDENCE OF the Descent of Man from Some Lower Form," the first chapter of *The Descent of Man*, his magnum opus on human evolution, Charles Darwin drew on evidence from comparative anatomy and embryology, revealing how similarities in the structure of living animals provided important clues to human evolution, through traces of common ancestry with other animals. He was able to make a convincing case without recourse to the fossil record—which for hominins* was virtually

---

* In this chapter and throughout this volume, the term "hominin" refers to human ancestors and extinct relatives. Hominins have small canine teeth and anatomical adaptations for upright walking and are more closely related to humans than to any of the other apes. The term "hominid" refers to a family of extinct and living primates colloquially called the "great apes." There are dozens of extinct hominids known from the fossil record. The living hominids are humans, chimpanzees, bonobos, gorillas, and orangutans.

nonexistent at the time—or indeed to genetics. Looking back from where we find ourselves today, with a specious hominid fossil record showing how our kind evolved in Africa over millions of years, and with our knowledge of genetics, it seems perhaps extraordinary that Darwin could have deduced so much from so little. And perhaps we have forgotten just how much evidence for evolution there really is buried in the bodies of living animals and embryos.

How did we get here? What's the evidence for our having evolved? The proposition that the human species arose through the same natural processes that produced every other species on this planet is testable. We don't just have to take it for granted: that's not how science works. We can test out this proposition, and the way we do that is by looking at the evidence.

Evolution, in the biological sense, is a change in organisms over time, generation by generation. It happens because of the variation in inherited characteristics that is present in any population and because some individuals end up having more offspring than others. The first question, then, is: Do humans vary in ways that can be passed on to their offspring? The answer is clear: of course we do. We're all very familiar with the fact that children tend to bear similarities to their parents, in all sorts of ways. Nineteenth-century scientists, including Darwin, could observe those patterns of inheritance, even if they didn't know about the mechanism—DNA.

A further clue to the evolvability of humans, Darwin argued, comes from the pattern of variation seen in species across the globe. Just as we expect children to look more like their parents than like any more distant relatives, we would expect certain characteristics to be more frequent in one population compared with another. We would also expect geographic clusters of those characteristics to exist—even with all the movement of people that has happened over the millennia. Other animal species exhibit geographic variation like this, sometimes so marked that distinct varieties or even subspecies may be identified. While Darwin goes further than biologists would today in recognizing different races of humans (see Chapter 7 of this volume), it is clearly true that certain traits in humans show geographic variation.

But the definition of evolution involves a change in frequency of in-herited characteristics (and, as we now understand, in particular genetic variants) over time. It is not enough that there *is* variation today among human populations; we must ask whether the patterns of variation have actually changed. Have humans, like every other life-form, been subject to natural selection? Do we see any evidence of beneficial variations spreading through ancestral populations and of disadvantageous traits being weeded out? The answer is yes, and we can see this evidence of evolution clearly in the fossil record and in our DNA.

In Darwin's day, the fossil record of human evolution was practi-cally nonexistent. Fast-forwarding to the twenty-first century, we now have bounteous fossil evidence for human evolution (see Chapter 4 of this volume), including around twenty known fossil hominin spe-cies, forming a six-million-year-old family tree of two-legged apes that includes our ancestors. And we can see how particular adaptations— from anatomical features that improve the efficiency of bipedalism to the expansion of brain size—emerged and took hold among our ancestors.

In the century and a half since the publication of *Descent of Man*, there have of course been great leaps forward in our understanding of the nature of inherited characteristics, with the discovery of DNA and ongoing research into the function of genes and the reading of entire genomes. We know that our genetic makeup influences our anatomy, physiology, and behavior—and that some traits are more tightly con-trolled by our genes than others. We see that we are, at a fundamental level, made of the same stuff as other animals and subject to the same biological processes. The more detail we're able to discern, the more similarities we see between ourselves and other forms of life. And we can also see evidence for natural selection written into our genomes— in the conserving of important genes, the promoting of the spread of advantageous mutations, and the weeding out of variants that might have compromised fertility in the past, for instance.

But even back in the nineteenth century, before biologists had worked out what the *stuff* of inheritance was, before all those hominin fossils were discovered, there was plenty of evidence for evolution, and

Darwin knew exactly where to find it: in anatomy. Similarities in the structure of adult bodies, and in embryos, hint at links between living animals—links that Darwin knew were explained by common ancestry.

Further clues could appear when things went wrong with development. Occasionally, a developmental anomaly would seem to hark back to an earlier stage of evolution, producing a "throwback," or atavism. Darwin was fascinated by these anomalies; their existence—even when their genetic basis was unknown—suggested that humans were subject to the same laws of inheritance and development as other animals. As well as anomalies, he was also interested in anatomical variants that seemed to be nonfunctional and to represent echoes of ancient ancestors. He called these "rudiments," whereas we tend to refer to them as "vestigial features" today. Some of the vestigial features he wrote about have turned out to be more functional than he suspected; others have proved to be novel characters rather than vestiges left over from earlier ancestors.

Although Darwin may have been wrong about a few details, his brief survey of comparative anatomy and embryology, forming the first chapter of *Descent*, stands up well today. The fact of our evolution is indeed written into our bodies—with some traces of earlier ancestors very obvious while others are more hidden away. What's astonishing about such clues is that they come from careful study of living organisms— and provide compelling evidence for evolution on their own, without drawing on fossils or genetics, which now provide us with independent, corroborating bodies of evidence.

Anatomical and Physiological Similarities between Humans and Other Animals

Biologists have realized for centuries that humans are fundamentally similar to other mammals, sharing the same basic body plan. We can look at a human skeleton and find corresponding bones—which are given the same names—in a monkey, a bat, or a seal. We find the same when we turn our attention to the muscles, nerves, blood vessels, and internal organs. Even the human brain—which is larger, compared with body size, than that of any other mammal (see Chapter 2 of this volume)—still appears to be a variation on a theme. The human brain is three times the size of that of our closest living relatives, chimpanzees. The human cerebral cortex—the outer layer, consisting of cell-dense gray matter—contains twice as many neurons as that of chimpanzees. And yet the general anatomy of these two very differently sized brains is strikingly similar, especially in the lobes and to a fair extent in the fissures and folds. There are fewer obvious similarities when we look at the even smaller brain of a more distant relative, the macaque. While the human brain is not just a scaled-up version of other primate brains, it is, as Darwin noted, evidently related to these brains. And that of course is the key to it: all these anatomical similarities represent relatedness. We are connected with these other animals through threads of common ancestry. We can take any animal and trace its genealogy back until we find a shared ancestor of both that species and our own.

These types of similarities, which reveal how closely we're related to other animals, are not just to be found in our bones, our blood, and our brains. Darwin believed that there was more evidence of common ancestry to be found in human susceptibility to infectious diseases that also affect other animals. Zoonoses are diseases that can cross the species boundary, leaping from other animals to humans. They include bacterial, viral, and fungal infections as well as parasitic infestations. Of all the transmissible diseases to which humans are vulnerable, over 60 percent are zoonotic. It seems reasonable to assume that diseases are more likely to spread successfully (for the diseases) from one species to another if those two species are closely related. And some very significant zoonotic infections are known to have jumped from other primates into humans, including HIV, malaria, and monkeypox. The devastating Ebola outbreak that killed more than 10,000 people in West Africa between 2014 and 2018 was initially thought to have come from great apes, but careful investigation revealed that bats were a more likely source. Bats and horses carry Hendra virus, and mice carry the Lassa virus. Humans contracted our own form of bovine spongiform encephalopathy (BSE) from cattle. Flu viruses are notoriously capable of spreading from birds and pigs into human populations. And mammalian carnivores also turn out to be a particularly important reservoir for zoonotic diseases. So, although humans certainly can and do catch infections from other primates, we can be very vulnerable to pathogens from much more distantly related species.

The problem with using zoonoses to study the relatedness of humans and other animals is that this adds another layer of complexity. After all, the defining characteristic of a successful zoonotic disease is that it *isn't* fussy—it is a generalist. These diseases aren't adapted to the immunological landscape of just one host; their success depends on their ability to survive and thrive in a range of quite different animals.

Darwin does point to some aspects of physiology that seem to provide evidence of common ancestry, and here he is on safer ground than with the infectious diseases. His examples are anecdotal: he writes about monkeys becoming intoxicated when plied with alcohol—and even suffering from hangovers like humans. He notes that humans heal by essentially the same process we see in other animals. The depth of knowledge about animal physiology has grown enormously since Darwin's day; now we can delve into any physiological aspect and find homologous processes taking place in the bodies of humans and other animals. In fact, our understanding of human physiology has been enhanced through studying and comparing physiological systems among animals. The functions of the cardiovascular, respiratory, endocrine, nervous, musculoskeletal, digestive, and reproductive systems appear as variations on a theme throughout the animal kingdom, just as the anatomical structures of bones, blood, and brains do. The variation reflects adaptations to different lifestyles, different ecological niches, but also reflects the evolutionary history of each species.

Darwin noted the "close correspondence in general structure, in the minute structure of the tissues, in chemical composition and in constitution" between humans and closely related animals.[2] At the time he was writing and observing, in the nineteenth century, many similarities or correspondences were visible to him, but a whole host were hidden and yet to be discovered. The detailed structure of that "chemical composition" was to be elucidated over the course of the twentieth century, and that work still engages biologists today. We understand now how the hemoglobin protein molecule of a human differs—in its amino acid sequence and its structure, as well as its affinity with oxygen—from that of the extinct woolly mammoth. We know now that the molecule that carries information from one generation to the next, which underpins any heritable characteristic, is deoxyribonucleic acid—DNA. We can see how this molecule differs in its sequence and its expression from one animal to the next, humans among them. Darwin said that it wasn't possible to exaggerate the similarity between the other great apes and humans. He would surely have been delighted to learn that the genetic sequence of humans is remarkably close to that of chimpanzees, with a 96 to 98 percent similarity, depending on how it is measured.

As we can see, there is a wealth of evidence of relatedness, of common ancestry, with other animals when we look closely at anatomy, physiology, and biochemistry. We can look at structure and function at the level of whole organisms and see similarities there. We can focus right down to the level of molecules within the cells of different species and once again see similarities in structure and function. And what is crucial to understand is that these deep veins of similarity don't merely reflect adaptations to similar lifestyles. Indeed, they often appear to conflict with those demands. Form and function appear to be strongly influenced or constrained by something else as well, and what Darwin saw very clearly is that the "something else" was common ancestry.

And signs of common ancestry appear even more clearly when we turn our gaze to look not at fully formed, adult organisms but at tiny, developing embryos.

Similarities in Embryos

Darwin had written about the striking similarities among vertebrate embryos in his 1859 book, *On the Origin of Species*. He illustrated his point with an anecdote about the famous Swiss American anatomist Louis Agassiz: "Having forgotten to [label] the embryo of some vertebrate animal, he cannot now tell whether it be that of a mammal, bird, or reptile."[3] Darwin realized that the resemblances between embryos of different species could provide important clues about the animals' evolutionary relationships—clues that later become obscured by the appearance of specific adaptions in adult animals. In a creationist view of biology, the similarities between embryos (and adults) represented an abstract connection between animals in the mind of a creator. Under the new evolutionary paradigm, those resemblances spoke of real, physical links between ancestors and descendants.

A brand-new human being starts at conception—as a fertilized egg or ovum. At ovulation, the egg, containing half the genetic material needed to make a human, bursts free from the ovary. The egg, about a tenth of a millimeter in size, possesses the largest diameter of any cell in the human body. As it leaves the ovary, it takes a cluster of smaller supporting cells with it, and the whole mass is picked up by the waving, fingerlike fimbriae fringing the open end of the oviduct. At coitus, sperm are deposited in the upper vagina, and to reach the egg, they must traverse the canal of the cervix, travel through the cavity of the uterus, and enter the correct oviduct. Only a few make it that far: of the 100 million that start the race in the vagina, fewer than 100 reach their destination, and—if conception occurs—only one will fertilize the egg.

This process of conception is very similar across all mammals, and the sex cells—the sperm and the egg—also look very similar. As the fertilized egg starts to divide and develop, first into a cluster of cells and then into a fluid-filled ball, there is nothing obviously human about the way it looks. By five weeks after conception, the human embryo has developed a circulatory system and a series of segments along its back. It has also just started to sprout tiny buds that will develop into limbs. At this stage, it is still virtually indistinguishable from other mammal embryos at a similar stage of development and even bears some strange similarities to fish embryos. A series of ridges, separated by clefts, in the neck of the human embryo looks strikingly similar to the earliest precursors of gills in a fish embryo; there are even arteries running through these ridges in the human embryo. In the developing fish, similarly

Fish   Salamander   Tortoise   Chick   Pig   Calf   Rabbit   Human   Early

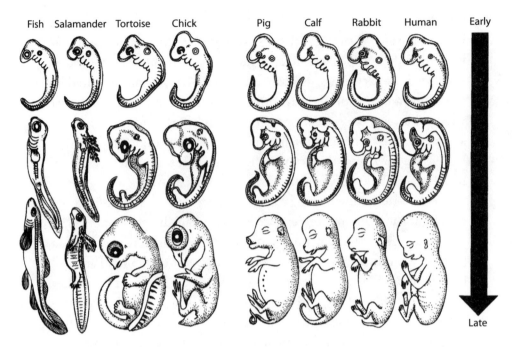

Late

FIGURE 1.1. Comparative embryology. (Redrawn by the author from Ernst Haeckel.)

positioned arteries will become the blood supply to the capillary beds of the gills, enabling the fish to pick up oxygen from the water it swims in. The equivalent arteries in the human embryo will never supply gills but instead get recycled and replumbed to form other essential blood vessels in the human head and neck.

The limbs of any vertebrate, whether a reptile, bird, or mammal, start as tiny buds in the embryo, which lengthen and grow, with digits forming at the very ends. Humans are no different to any other vertebrate in the way in which their arms and legs start to sprout and develop (Figure 1.1). But as embryonic development progresses, characteristic differences start to emerge between species. The more closely related the species, the longer it tends to take for those differences to emerge. At between five and eight weeks of development, it becomes easy to tell apart a human embryo from a dog embryo—while during this time the human embryo remains almost indistinguishable from a chimpanzee embryo.

It's not just the overall appearance of the early human embryo that is similar to other animal embryos. As particular structures develop in the human embryo, they can look very much like the equivalent but simpler structures in other, more distantly related animals. We see such striking similarities in the developing heart, kidney, tail, brain—and yolk sac.

The embryonic human heart starts off as two separate tubes, which then fuse together to make a single tube. This primitive heart tube, which is already contracting to push blood through it just four weeks after conception, begins to twist into an S shape. At this point, it looks strikingly similar to an adult fish heart. This similarity is then lost as the heart continues to fold and divisions form between its chambers inside.

In all vertebrate embryos, including humans, a structure nestling in the back of the abdominal cavity, called the mesonephros, acts as a kidney, filtering blood and producing urine. In fish and amphibians, the mesonephros persists as part of the mature kidney. But in reptiles, birds, and mammals, another embryonic structure—the metanephros— takes over to become the definitive kidney. The obsolete mesonephros gets recycled into other structures, including tiny tubes inside the testis in a male.

At five weeks, the human embryo possesses a tail—something retained by many other mammals, of course, into adulthood but practically lost in humans, reduced to just four tiny, fused vertebrae at the tail end of the spine: the coccyx. At an early stage of development, the human brain looks simpler and quite similar to the brains of adult monkeys.

Human embryos possess a curious structure that lies outside the actual body of the embryo—a yolk sac. In egg-laying animals, this yolk sac is stuffed with nutrients and is an essential part of the embryo's life-support system. In placental mammals, including humans, the embryo stays inside its mother's body for much longer, and the placenta facilitates the transfer of nutrients from the mother's blood to the developing embryo. In an early human embryo the yolk sac is empty and very small compared with the yolk of a bird embryo. The human embryo's yolk sac is eventually obliterated by the swelling amniotic cavity. It is effectively obsolete—yet it still develops. It's a memory of distant egg-laying relatives that has yet to be erased; perhaps it's simply too difficult to write it out of the program of development.

The correspondence between human embryos and the adult forms of other animals that might be considered to be simpler and more primitive led to an idea about how organisms evolved over time. This idea, called "recapitulation," was popularized by the nineteenth-century German biologist Ernst Haeckel, who believed that evolutionary change occurred through new modifications being added on at the end of embryological development. Recapitulation meant that the embryological development of an organism would reflect the exact sequence of its evolutionary development. A human embryo, for instance, might be expected to pass through stages when it looked like a fish, an amphibian, and then a reptile before it started to look like a mammal. But by the turn of the twentieth century, experimental embryology and genetics had comprehensively demolished the theory of recapitulation. Embryological development was more complicated than the theory suggested, and genetic changes could kick in and alter the program of development at any point. The crucial ideas behind recapitulation—that extra features could be added only *at the end* of embryonic development, and that embryos passed through stages equivalent to the progression of *adult* ancestors—no longer stood up to scrutiny.

Even though we now know recapitulation to be wrong, there *are* links between embryological development and evolutionary history. Haeckel was mistaken—animals don't have the equivalent of their adult ancestors telescoped into their embryos. But Darwin realized that the resemblances between embryos were still meaningful. They revealed important clues about evolutionary relationships between animals—clues that were often lost or obscured in adult animals.

Vestigial Features

Darwin used comparative anatomy, physiology, and embryology to build a compelling case that humans were evolved creatures and related to every other animal species. Patterns of similarity and difference could reveal just how closely related humans were to a particular species—they were closer to any other ape, for instance, than to a dog. But there was one type of similarity in body structure that Darwin found particularly fascinating and that he presented as the "smoking gun" evidence that humans had evolved. This evidence came in the form of odd anatomical parts that appeared to have little or no current function. These parts couldn't be explained as anything other than hang-overs, remnants, vestiges, or "rudiments," as Darwin called them, of earlier stages of evolution.

The trajectory of evolutionary development isn't always toward increased complexity. Some aspects of anatomy may become simpler over time, especially if they fall out of use or begin to present a disadvantage to an organism. Darwin realized a need to distinguish between anatomical parts that appeared to be vestiges of once functional structures and "nascent" parts that were in fact very useful and perhaps even likely to become more complex over time. The incisor teeth of ruminants, which never fully erupt through the gums, may be considered to be vestigial—they are on the way out, rather than on the way in. Vestigial structures also tend to be very variable in a population—there's little or no selection pressure acting to keep them a particular way while they still exist at all. Often, of course, such structures may disappear entirely, but they can sometimes be resurrected, if—as we now know—the genes to produce them still exist. Some essentially vestigial structures have found new uses; the wings of penguins are no longer useful for flying but perform admirably to propel the birds underwater.

If an anatomical structure falls out of use completely, it is likely to disappear over time. Its disappearance is hastened if it causes a disadvantage to the organism, in which case natural selection will favor variants that lack the particular characteristic or indeed possess a smaller version of it. The disadvantage could be purely energetic—it's costly to grow and maintain body parts that are no longer useful. But Darwin was perplexed as to how natural selection could completely eliminate a structure. Once the structure was too small to be particularly costly, the selection pressure against it would surely be minimal. He was left at a loss because he didn't know the precise mechanism of inheritance. But we now understand the influence of processes other than natural selection in evolutionary change and especially the role that chance plays. Particularly when there is no selection pressure to maintain something, genetic drift can mean that features that are no longer useful are quickly lost through chance affecting which individuals survive and reproduce. On the other hand, it may prove overwhelmingly difficult to eliminate a particular feature. The molecular control of development is complicated, and tinkering with the program of embryological development is fraught with risk.

The human body is littered with anatomical oddities that seem to play at most a very minor role today but represent remnants of structures that were once useful and larger or more widespread in the bodies of our ancestors. Many mammals possess muscle in their subcutaneous tissues, often forming a near-continuous sheet of muscles over the thorax, abdomen, and upper parts of the hind limbs. This muscle sheet is known as the panniculus carnosus; we see it in action when a horse twitches its skin to dislodge an annoying fly, or a dog shakes its skin to throw off water droplets. Other possible functions of the panniculus include shivering to produce heat, wound closure, and protection against abrasion. In humans, remnants of this muscle sheet are present at just a few locations, and they are variable between individuals.

The muscles embedded in the scalp, in the forehead, and at the back of the head are part of this superficial group, as are tiny muscles around the ear. Another surviving component of the panniculus is the extremely thin platysma muscle, whose fibers radiate out across the front of the neck just under the skin. Many of the surviving muscles are still useful in humans. Darwin presents a brilliant anecdote about a French family whose members could all use their scalp muscles to throw books off their heads, but in fact those scalp muscles are more generally useful in facial expressions, as are other muscles underlying the skin of the face itself. Other useful, persistent elements of the panniculus include the pectoral and trapezius muscles, which move the shoulder and arm.

Of the preserved parts that we find in humans today, the most variable seem to be the muscles that are least useful under current conditions. These muscles are often absent, and it doesn't seem to matter to the individual that they're missing. The sternalis muscle—a strange strip of muscle fibers over the breastbone—is found in only about 4 percent of people.[4] The weedy palmaris longus muscle in the forearm is often missing, and at different rates in different parts of the world.[5] A study in China found it to be absent in around one in twenty people, whereas a Turkish study found it missing in one in three people. The palmaris longus has recently been found to be useful in a new way: plastic surgeons harvest the tendon to use in reconstructive surgery on the hand, but this is very unlikely to drive up its occurrence.

It is much harder to argue for any current role for the three tiny auricular muscles around the external ear, which are surely just remnants of more developed muscles in distant mammal ancestors that could still move their ears toward sounds or prick them up when alert. As well as those small auricular muscles, the entire auricle of the external ear is often thought to be vestigial. Darwin thought so, writing in *Descent*: "The whole external shell of the ear may be considered to be a rudiment."[6] But actually, the auricle proves more useful and important than Darwin suggested; in a study in which the grooves in the outer ear—not the ear canal itself—were filled in with silicone, neuroscientists discovered the owner's ability to pinpoint where sounds were coming from was reduced. A smallish, shell-like auricle is not something that's uniquely human—other apes and monkeys have very similar external ears. Rather than a vestigial structure that's effectively shrunk and become less mobile through disuse, then, it may be that monkey and ape ears represent an adaptation to localizing sounds in three-dimensional space—something that's very useful for an arboreal, social primate.

Darwin picked up on an anatomical variant in the auricle, though, that really does seem to be a vestigial structure. On the folded-over outer lip of the ear—known technically as the helix—some people possess a small, forward-projecting point. Although it was the English sculptor Thomas Woolner who first described this anatomical variant and speculated that it could be atavistic, it has become known as "Darwin's tubercle." Sometimes these tubercles even stick out slightly, so they can be seen from behind. Recent studies have reported widely different rates of occurrence of Darwin's tubercle, variously finding it present in around 10 percent of Spanish adults, 40 percent of Indian adults, and 60 percent of Swedish children.[7] Darwin drew attention to the fact that some monkeys, such as baboons and some macaques, tend to have slightly pointed ears. His assertion was that the entirely benign tubercle seen in some human ears represented the remnant of a pointed ear of an earlier primate ancestor, and he may have been right about this.

Another small, vestigial structure noted by Darwin appears at the inner corner of the human eye: the semilunar fold. This appears to be a much-reduced version of the well-developed "third eyelid," or nictitating membrane, which is seen in some fish, amphibians, reptiles, birds, and a few mammals, including marsupials, seals, polar bears, and camels.

Darwin considered the human sense of smell to be reduced as far as to be essentially rudimentary and vestigial. The nineteenth-century neuroanatomist Paul Broca had already classified humans as "non-smellers," drawing attention to the small size of the olfactory bulb—the swelling at the end of the olfactory bundle, or tract of nerves, that lies tucked under the frontal lobe of the brain.

The importance of smell to most mammals is evident from just how many genes they possess for olfactory receptors—the proteins that "recognize" odor molecules: more than a thousand. Humans seem to have a respectable number, around 800. But, in fact, around half of these are inactive and have mutated to the point that they can no longer be "read" and translated into proteins. Mammals ranging from mice to dogs have three times as many active smell genes as humans.

When did our ancestors start to lose their active smell genes? It used to be thought that this disinvestment in the sense of smell happened alongside an improvement in the visual system. Lemurs, lorises, and monkeys from the Americas have just two types of color receptor in their eyes; Old World monkeys and apes have three, giving them so-called trichromatic color vision. It was hypothesized that the disinvestment in smell happened among those primates that developed color vision. It's a neat story, but more detailed analyses of entire genomes have failed to support it. Instead, the loss of smell genes seems to be a very general tendency among primates. Most primates appear to have only 300 to 400 active smell genes. In fact, although we tend to think of ourselves as particularly lacking in the olfactory department, humans possess more active smell genes than chimpanzees, orangutans, marmosets, macaques, and bush babies.

The story of smell doesn't end with the number of active smell genes and the size of the olfactory bulb. Humans may have fewer active smell genes than rodents and a relatively smaller olfactory bulb, but they have more complex areas of the brain with which to process the incoming information. When making comparisons between humans and other animals, the range of smells tested makes a huge difference. Using an appropriate range of smells, humans can do better even than dogs and rodents.[8] Darwin, following Broca's cue, may have been wrong about the olfactory limitations of humans, but he was right about strong links between the sense of smell and memory: a particular odor can bring back memories and evoke strong emotions.

Darwin also commented on humans' apparent nakedness—certainly one way in which we appear to be markedly different from other primates. He drew attention to the fact that there is enormous variation in just how hairy people are within any given population. This variability is again what would be expected of a largely vestigial characteristic that hasn't been under strong selection. Even when people appear to be completely naked, they do of course possess hairs—though those hairs are extremely fine. In fact, humans possess just as many hair follicles per square centimeter as other great apes. Darwin wrote, "There can be little doubt that the hairs thus scattered over the body are the rudiments of the uniform hairy coat of the lower animals."[9] He was right; there can be little doubt that humans have evolved from a hairier ancestral state, but the fine hair that we have retained is far from useless. Human hairs may provide little in the way of insulation or coloring, but they turn out to be extraordinarily useful for sensing the presence of a crawling ectoparasite, such as a bedbug, tick, or louse. Perhaps this explains why human hair has not been lost completely.

Darwin noted that the fetus is covered with quite a thick coat of hair, known as lanugo, during the sixth month of development. This fetal covering, he posited, must be the remnant of fur that would have been retained up to birth in hairier ancestors. And, indeed, chimpanzee fetuses also develop lanugo, and rather than losing this fur, as the human fetus does, they are still covered in it at birth. Human lanugo appears to be a genuine vestigial trait, then: it is there just because we had ancestors who were once much more thickly furred than we are today.

Wisdom teeth—the last molars to erupt in our jaws—are also candidates for practically vestigial features. Some people congenitally lack these third molars; others have troublesome wisdom teeth extracted with no apparent downside to this reduction in dentition. Humans are not alone in exhibiting variably missing molars; other mammals often have absent teeth. Placental mammals have tended to progressively lose teeth over the course of evolutionary time, and scientists have begun to uncover the genetic and embryological basis of these changes in dentition. Experimental models in mice have shown that subtle changes to levels of signaling molecules during development can have widespread effects, from complete tooth loss to altered shape of teeth. This means that, while tooth pattern may be fairly stable within a species, there is some plasticity that may facilitate swift adaptation to different environments and diets. In particular, molars develop in an anterior to posterior sequence, and the first to develop exert an inhibitory influence on subsequent molars. If jaws become smaller, the inhibitory effect on the last molars—in humans, the wisdom teeth—is likely to be more extreme.

The mouth is, of course, the entrance to the alimentary canal, and some way down this tract—a few meters down, in fact—lies a structure that is perhaps more famous than any other for being vestigial: the appendix. Once thought to be unique to humans and the other apes, the appendix—a narrow diverticulum off the cecum, at the commencement of the large intestine—is now known to be a widespread feature of primates, present in lorises and lemurs, various monkeys, and apes.

In some mammals—herbivores that employ hindgut fermentation—the equivalent part of the gut appears to be a huge diverticulum, which, rather confusingly, is also known as the cecum. This pouch is essentially a fermentation vat, in which friendly bacteria break down cellulose into sugars that can be absorbed. In humans and other apes, the appendix is far too small to be a useful site for fermentation. It's often touted as useless, and Darwin cited its variability as support for the appendix being vestigial and essentially nonfunctional.

Although small and highly variable, the appendix has long been known to contain a high concentration of lymphoid tissue and has been regarded as a useful, though clearly nonessential, part of the immune system. However, it is now thought that the appendix may also play a very important role after bouts of gut infection and diarrhea, as a "safe house" for beneficial gut flora, which can then quickly recolonize the intestine.[10] Finally, to be a vestigial structure, it would also need to be shown that the human appendix had evolved from a much larger version in a remote ancestor. This is not clearly the case, and in fact it appears that the appendix is instead a new, so-called derived character in some primates—not a vestige at all, then, but a novel product of evolution.

Darwin also focused on a number of anatomical variants in the human humerus, which appeared to reflect more common anatomical features in other animals. One of these was the supracondylar spur, found in lemurs and cats, which provides extra area for the attachment of pronator teres muscle and under which the median nerve and brachial artery pass. An apparently homologous feature—the supracondylar spur and accompanying ligament of Struthers—occurs very rarely in humans and can cause painful compression of the median nerve.[11] This bony spur does appear to be some form of throwback.

Atavisms of this type occasionally appear in humans and are often picked up because they create problems. Another example relates to tendons running through the hand to the fingers, which in rare cases have small associated muscles in the palm that can be painful. Although the genetic basis of these anomalous muscles isn't understood, they appear to be equivalent to the small flexor muscles found in the hands of limbed reptiles and amphibians.[12]

Darwin also presented the human coccyx as an example of a vestige. It is clearly the reduced remnant of a tail, now comprising just three to five vertebrae. But in fact, it is functional in that it provides an important site of attachment for the pelvic floor muscles. On the other hand, the associated coccygeus muscle in humans, which lies on the internal surface of the sacrospinous ligament, is entirely obsolete. But its position suggests it really is the vestigial remnant of once-useful tail muscles.

Darwin extended his tour of potential vestigial structures to examine differences within our species between the sexes. Darwin points to the rudimentary mammae of male mammals, including humans, and to the small prostatic utricle in the male, which has the same embryonic origin as the uterus in the female.

Human embryos start off with a unisex kit of parts: a pair of gonads and two pairs of tubes, the Wolffian and Müllerian ducts. At six weeks after conception, the gonads and the ducts associated with them are identical in both male and female embryos. But then, as development continues, differences begin to appear. In female embryos, the Wolffian ducts all but disappear, and the Müllerian ducts are retained. Their lower ends fuse together to form the uterus and vagina, while their uppermost portions remain separate, forming the oviducts. In a male embryo, it's the other way around: the Müllerian ducts regress, while the Wolffian ducts stick around, forming a vas deferens on each side. But in both sexes, there are tiny remnants of the pair of ducts that regress. In women, part of the Wolffian duct may persist close to the ovary; this remnant is known as the epoophoron. In men, the prostatic utricle—a small outpouching of the urethra where it passes through the prostate gland—is indeed a remnant of the Müllerian ducts, as Darwin noted. There's also a small bump, the appendix of the testis, that represents another remnant of those ducts in a male.

These quirks of embryonic development were known about in the nineteenth century, but now we also understand something about the molecular mechanisms that underlie them. In mammals—including humans, of course—determination of sex depends on chromosomes. Females generally possess two X chromosomes, and males possess one X and one Y chromosome. There are exceptions to the rule, as there always invariably are in biology. In rare cases, an individual may be born with a single X chromosome or with three or more sex chromosomes. The links between genotype and the development of phenotype are also very complex, involving cascades of hormones and interactions between hormones and receptors in different tissues. While biological sex in humans can be considered to be generally binary, developmental biology shows it is not exclusively, absolutely binary.

Most individuals possessing XY sex chromosomes will develop as males. The Y chromosome contains a gene called SRY—for the "sex-determining region of the Y chromosome"—which works in concert with other genes to prompt differentiation of the testes. Inside each developing testis, certain cells then begin to produce a substance known as anti-Müllerian hormone, or AMH. This hormone spells the end for the Müllerian ducts in male embryos. Other cells in the testis begin to produce the hormone testosterone, which virilizes the Wolffian ducts, urging them to develop into the vas deferens. Testosterone also masculinizes the developing external genitals. But female development is far from a default setting in the absence of a Y chromosome—it's also an active, gene-directed process. At least three genes essential to ovary development have been discovered, all of which inhibit "male" genes and lead to a feminization of genitals. Further downstream in the process, hormones produced by the ovary encourage the development of the other female reproductive organs and feminize the external genitals.

In some ways, Darwin seems to have wandered off on an intellectually interesting but potentially distracting diversion when he writes about vestigial features in each sex. After all, his main aim is to lay out the evidence for humans as an evolved, not created, species. But of course, for creationists, males and females were also separately created—and there should be no reason the human male and female should each bear vestiges of the other's organs. The fact that they do hints strongly at underlying mechanisms and constraints of development and inheritance. The appearance of vestiges among the sexes implies mutability and the existence of a biological mechanism by which it would be perfectly possible to modify structures over time.

The Bearing of These Facts on the Origin of Humans

When Darwin wrote *Descent of Man*, he was able to pull together evidence for humans as evolved organisms from his wide appreciation of comparative anatomy and embryology. Those comparisons still stand, even if he was wrong about a few structures that he thought were vestigial. In the century and a half since the publication of his book, many more similarities in structures have been uncovered, and of course we now have the potential to make comparisons at a minute level: at the scale of molecules. DNA sequencing has revealed a deep vein of similarity connecting species that appear completely different as adults or even as embryos. A fruit fly is a very different organism from a human, and yet the Hox genes that determine patterning in the fly embryo—telling segments at the front end to become the head and segments at the back to become the tail—are almost identical to the equivalent patterning genes in a human.[13] Everywhere we look, we find evidence for connections between living organisms—from macroscopic anatomy down to the molecular sequences of proteins and genes.

Darwin was not the first to notice similarities between humans and other animals, but he was the first to be emphatic about their meaning: homologies reveal shared ancestry. "The homological construction," he wrote, "is intelligible, if we admit . . . descent from a common progenitor."[14] Richard Owen, the first curator of the Natural History Museum in London, had published a book in 1849, On the Nature of Limbs, in which he carefully compared mammal skeletons.[15] Owen remarked on the similarities between the forelimb bones in a horse, mole, dugong, and human and concluded that each one represented some sort of variation on a theme around a mammal archetype. Darwin firmly rejected this conclusion, writing, "It is no scientific explanation to assert that [mammals] have all been formed on the same ideal plan."[16] The explanation for the similarities wasn't an abstract ideal or archetype but something very real indeed: shared ancestry, or as Darwin put it, "community of descent."[17]

Darwin also knew that these patterns of similarity extended across the animal kingdom much more generally. And now that we have biochemical and molecular data to add to the picture, we can see that these patterns extend even further, revealing deep shared ancestry between animals, plants, fungi, bacteria, and archaea—all of which use the same molecules of inheritance: nucleic acids. These patterns of similarity—anatomical, embryological, genetic—reveal how every life-form on this planet is part of a great, branching family tree of life. Closely related species, with a more recent common ancestor, share more similarities than distantly related life-forms. Detailed surveys of similarities and differences, particularly genetic,[18] can be used to reconstruct phylogeny, revealing the family tree of all life on earth. Knowing rates of mutations in genes, the branch points within the tree—where one species began to diverge from another—can even be dated. Even without a fossil record, then, we can see evidence for evolution having happened: in the bodies, embryos and genes of living organisms.[19]

Our species, *Homo sapiens*, is just one tiny twig on the great tree of life. Humans—like every other species—are a product of evolution. As Darwin put it, it is only our natural prejudice and arrogance that has ever stood in the way of accepting the overwhelming evidence of our own evolved nature.

2

Remarkable but Not Extraordinary: The Evolution of the Human Brain

Suzana Herculano-Houzel

My object in this chapter is solely to shew that there is no fundamental difference between man and higher mammals in their mental faculties.

CHARLES DARWIN, *THE DESCENT OF MAN*, CHAPTER 2[1]

IN 1871, twelve years after making waves in science and society at large with *On the Origin of Species*, Darwin visited, in *The Descent of Man*, a lingering question his previous work had opened, only to leave unanswered: How far were his general conclusions about the descent of species through modification applicable to humans, if at all?

His start to Chapter 2, "Comparison of the Mental Powers of Man and the Lower Animals," left no doubts that his answer was "all the way." The putative mental continuity between humans and other species was a piece of evidence as fundamental to establishing Darwin's case for human evolution as was the far easier to demonstrate physical continuity. But in 1871, there was not much to argue for (or against) any similarity between human and nonhuman behavior, much less between the neurobiological underpinnings of mental capacities of humans and other species.

If only Darwin could have known in the late nineteenth century what a then newly born science, neuroscience, would reveal over the coming 150 years. Neuroscience originated with anatomists, physiologists, and physicians who were already at work at the time of Darwin's writing of *Descent*, and who over the next few decades became more and more interested in systematically tracing the makings of the mind to the workings of the brain.

In 1863, French physician Paul Broca had described, before members of the Anthropological Society of Paris, how language was disrupted by lesions to the frontal lobe on the left side of the human brain. That confirmed Prussian anatomist Franz Gall's views that different parts of the brain had different functions in generating the mind[2]—much as the various organs of the body were already known to have separate, if interrelated, roles in supporting life.

It is easy to understand how, compared with some obvious similarities between humans and other species in bodily shape and overall organization, the mental powers characteristic of humans would appear strikingly different from those of other species and thus cast doubt on any proposition that humans descended from some "lower form."[3] Descent presupposes gradation, and by the late 1800s, any transition between an apelike form and the modern European men carrying out such studies seemed outlandishly sudden, not to say impossible. There were too many "missing links," and not just in the fossil record. How could the seemingly extraordinary human intellectual faculties possibly arise from a brain so like those of so many other species, only larger?

Curiously, over most of the 150 years since Darwin's *Descent*, neuroscience has addressed that fundamental question of how mammalian brains are generated and organized in *similar* ways across species, from mouse to human, while simultaneously harnessing expectations and theories of human exceptionality. Evolutionary conservation across mammalian brains and bodies—the very foundation that allows clinical and neuroscience research to use mice, rats, cats, rabbits, dogs, pigs, and monkeys as laboratory surrogates for humans (and to some extent even chicks, fruit flies, and worms)—has coexisted with the suspicion that deep, deep down, the human species is different where it matters the most: its brain. Darwin would have been befuddled, I think, at how his legacy could at the same time be embraced (humans evolved, like any other species) and disputed (for surely human evolution must be distinguished by some unique set of events and features)—and by the same individual scientists.

Of course, if the human foreigner is easily and unmistakingly singled out in a band of our closest living relatives, chimpanzees, and not just because she is the only one cloaked in fabric, it is because various features *do* differ between the species. But that was exactly Darwin's point: that over time and generations, species that once were one and the same retain some similarities while still becoming more and more different in other ways. The question that mattered for Darwin in Chapter 2 of his *Descent of Man* was not whether human mental faculties belonged with the former (the similar) or the latter (the different) but whether human intellectual faculties, clearly "superior" to those of "lower animals,"[4] differed only in degree from the intellectual faculties of other species, or also in kind. It is a question that would have remained restricted to metaphysics and science fiction—for how could one get into another animal's mind to know it?—if not for the cunning and persistence of a legion of scientists.

I'd like to make clear that, Darwin's arguments notwithstanding, the exact answer to the question of whether human brains and mental functions differ in degree or in kind from those of other animals is actually irrelevant for the purposes of establishing whether or not modern humans descend from some relative shared with great apes. That has by now been amply and solidly determined by decades of work by paleontologists, archaeologists, primatologists, geneticists, biochemists, and biologists. So much accumulated evidence makes it today an undisputed fact that humans in their modern incarnation—size, shape, and all—did not exist on the planet as recently as half a million years ago, just as it is known that splendorous *Triceratops* and *Velociraptor* once roamed the Earth but now only modest-size birds serve as reminders of their existence. What persists as *hypothesis* is the narrative of precisely *how* that happened.

Still, it is an interesting point: whether human brains and therefore their minds differ from those of other primates, and even other mammals, in kind or simply in degree. It is an important one, too, and for more than philosophical reasons or self-serving hubris. If the human brain is essentially made in the image of other primate brains, as Darwin submitted, then much insight is to be gained into our own humanity from studying and understanding other fellow primates and their ways and behaviors. If the human brain is mostly just a large primate brain, and we understand how a primate brain is both similar and different from a rodent brain, then there is plenty to be learned by studying the brains of mice and rats, which develop in three weeks, not forty, and then mature, age, and die in the tractable space of two years, not eighty, which makes much of human-relevant research feasible within the lifetime of individual researchers. And once mental functions can be mapped onto brain circuits, then understanding the evolution of the mind becomes a much less ethereal task, for reconstructing the evolutionary history of the brain can offer direct insights into what mental functions are or were available to different brains, living or dead.

Neuroscience has learned much in these 150 years about how the brain is organized and how it generates the mind, and these insights have informed the comparison between the "intellectual faculties"[5] of humans and other species. Whereas it is likely that the brains of insects, mollusks, and vertebrates appeared independently in evolution (even if organized by many of the same genes), all vertebrates and their brains are solidly rooted together, forming one large genealogical tree that descends from the same common and exclusive pool of ancestors. I will stick to the brains of vertebrates here, although, as it turns out, many of the most basic principles of brain structure and function are applicable across the board: organization in loops, spontaneous activity modulated by the senses, and associative connections that create complexity, flexibility, and the ability to reference the self.

The first principle is that all nervous systems, in all animals with one such system, are organized as closed loops that feed signals from the body to the brain and take signals from the brain back to the rest of the body. The former are conveyed through the senses; the latter, through nerves to muscles and viscerae, including glands, which make changes to the body—changes that then immediately circle back to the brain through the senses. The important part about the loop arrangement is that there is no beginning, middle, or end to signal trafficking in the nervous system. Every sensation can lead to an action; every action leads to a sensation. The loop becomes active as it forms and only silences in death.

Reflex movements, recognized in the nineteenth century,[6] are a type of action that occurs through the simplest of loops: those formed directly between a sensory surface and the muscles that move it. The loop is short and simple, and reflex actions are so named because they quickly and stereotypically "bounce back" from the nervous system directly or indirectly to the body part of origin—a finger being pulled back after touching a hot stove, for example. If there is a triggering stimulus, the reflex action always occurs and always in the same way.

Reflexes were already known to scientists and physicians in Darwin's time and were recognized as "basic units" of behavior. Herbert Spencer, and then Charles Sherrington, famously proposed that behavior was a series of coordinated reflex actions chained together, one right after the other.[7] This was an influential view that steered many lines of research well into the late twentieth century, until two discoveries rocked the field. One was Sten Grillner's demonstration that neurons form circuits that can generate patterned behavior, such as rhythmic swimming, walking, or breathing, but that patterned behavior could happen in the absence of patterned sensory input.[8] All it takes is a steady signal that turns the entire circuit on. Increasing or decreasing the strength of that steady signal suffices to modulate the behavior, turning a walk, for instance, into a trot and then into a gallop.

Even more transformative was the growing recognition in the 1990s that the brain is dominated by "ongoing activity": spontaneous variation in how excitable different neurons are at any moment, variation that suffices to turn circuits on and off independently of sensory input.[9] The ongoing, internally generated activity can spread among neurons and circuits, and take on patterns in space and time that mimic those elicited by the senses. For instance, for a person with eyes closed in a dark room, the image of a loved one on a sunny or rainy day can come to mind. Such internal activation of sensory representation constitutes the basis of memory recall, imagination, and dreaming.[10] Starting in the late twentieth century, then, there was growing recognition that animals and their brains are not necessarily confined to their present environment and circumstances. Neurons were upgraded from mere transistors in a circuit to autonomous units that could generate patterns of their own. Sensory input from the environment was downgraded from master controller of behavior to a modulator that interfered with the ongoing activity of the brain and thus nudged or even pushed behavior this way or that way but was neither necessary nor sufficient to cause behavior, except for those truly reflexive actions. In this scenario, mental activity, however defined, must be something that brains can generate spontaneously, depending on how their loops are organized, and to a degree that is in some proportion to how complex these loops are. And if it turned out that those neuronal loops and circuits in the brain were organized in similar ways across species, then the inevitable conclusion would be that the minds that those brains can generate must not differ terribly in kind, just as Darwin hypothesized.

As it turns out, vertebrate brains are much more similar to than they are different from one another in how their neuronal loops are organized into circuitry. All of them have simple, direct loops between body and hindbrain or spinal cord that suffice to generate reflex actions, such as adjustments in blood pressure, blood flow to the organs, automatic movements of the eyes that track objects, or small corrective movements that maintain body posture. Those loops are connected to circuits in the hindbrain—the part of the brain connected to the spinal cord—that integrate sensory information and ongoing activity and feed it back onto the outgoing neurons in the lower loops. These hindbrain circuits are the unsung heroes of behavior, those structures that effectively, ultimately, control whether one is awake, asleep, or somewhere in between; when one starts walking, breaks into a gallop, or stands still; when breathing stops and swallowing, talking, or singing happens instead; where eyes and ears are directed; how much blood pressure rises or drops; what digestive enzymes are secreted and in what amounts.

These hindbrain loops, in turn, are subject to the influence of yet other loops—ones they establish with the cerebral cortex. The cerebral cortex is a structure formed almost all the way up in the neural tube that gives rise to the nervous system in early development, posterior only to the part of the brain that controls all glands in the body: the hypothalamus. Just as the hypothalamus is the master of all things secreted in the body, the cerebral cortex, by virtue of its connections with all other loops that operate the body, has control, or at least influence, over all actions of the nervous system that are executed by those lower loops. Even the cerebral cortex is subject to modulation by other loops, such as the basal nuclei and cerebellum, which make actions smooth and fluid and keep sensory surfaces stable and ready for what is expected to come.

Within the cortex itself there are even more loops, such as those that bring information together from all the senses into the hippocampus and back to the cortex. New memories are formed as the hippocampus weaves together sensory associations that become written into the circuitry of the brain. Without memory, every animal would be forever stuck in the present, doomed to reacting to stimuli and asking the same questions over and over again. With memories, animals gain a past, and with the ability to evoke the past to make plans, they gain a future. Other parts of the cerebral cortex, especially the prefrontal loops, generate representations of those plans and associate them with information from the senses on the current context, position, and situation of the individual. With these associative cortices, behavior gains complexity. A key function of these particular cortical loops, for instance, is executive control: the capacity of going beyond simply responding to the present reality as assessed through the senses by modulating behavior according to internal reasoning and plans. Yet other cortical loops form the default-mode network, which allows referencing actions and sensations to a self anchored in the body, and the social network, which allows the cortex to represent others and their feelings and intentions, and thus allow behavior of the individual to take others into consideration.

The cerebral cortex, in control of all other loops in the brain and endowed with its own internal loops, thus sits in a position that allows it to influence behavior by generating past and future, internal plans, preferences, and goals, and representing the self and others in a spatial, temporal, and cognitive context. A cerebral cortex is not strictly required for an animal to behave (in the purest sense of doing observable actions). The hindbrain handles that very well, and even human babies born without a cortex will nurse, cry, and smile. But when a cerebral cortex exists, it allows for a new level of integration of information that no hindbrain alone attains, with behavior that can be flexible and complex—never the same twice, never fully predictable from the senses.

All vertebrates possess a cerebral cortex to some degree, whether large and obvious (as in humans) or very discreet (as in fish and reptiles), layered (as in mammals) or arranged in compact chunks (as in birds). Where it has been examined, functional cortical connectivity— the layout of what structures are linked and work together—turns out to be very similar across species. Humans, like monkeys, rats, mice, and even pigeons, have cortical loops that receive copies of all information processed by the hindbrain, harbor information-processing hubs in a recognizable hippocampus and associative cortical areas, and function in tight loops with basal nuclei and the cerebellum.[11] Monkeys and even rats, like humans, have an anatomically and functionally recognizable self-reference network of neurons whose coherent activity supports self-awareness, anchoring the self to the body and its location. In all three species, coherent activity in this network dissolves in sleep and anesthesia.[12] Along with it goes human consciousness;[13] it is only reasonable to infer that sleeping or anesthetized mice and monkeys are unconscious for similar reasons.

There remains little to be disputed over whether brains of humans and other vertebrate species are organized in similar ways. They are, inasmuch as the same structures are recognizable and connected in similar ways. The questions that remain are questions of degree—a reality of modern neuroscience consistent with Darwin's predictions. One is: to what degree are functional brain networks really similar across humans and other species? Would their neurons be interchangeable across species? (To the extent that human neurons survive in mouse brains, we do know that neurons can be interchanged).[14] Is a pyramidal neuron in, say, the human insular cortex set up with the same ion channels, neurotransmitter receptors, and second messenger cascades as a pyramidal neuron in the mouse cortex? (Remarkably, yes: neuronal cell types are defined by very similar constellations of genes expressed in mouse and human).[15]

We know to expect discrete genetic differences across species—one amino acid here, maybe a small stretch over there—given the well over 50 million years that separate the last common ancestor shared by mouse and human. Such small differences may produce a small effect, none whatsoever, or maybe completely rewrite the functioning of a cell. A collection of such small differences certainly lies behind the obvious differences in appearance between humans and chimpanzees, which share an ancestor that lived only 6–7 million years ago. It seems only logical that similar small differences exist between their brains and neurons.

Indeed, the past decade has seen a surge in studies that seek and find differences between human genes and their counterparts in other species, especially in genes expressed in the brain. Curiously, the idea that the human brain must be fundamentally different from others is contrary to Darwin's reasoning and resonates instead with expectations of his contemporary Alfred Russel Wallace. But so far, no striking differences have been found that would cause radical changes of *quality* rather than gradual differences of degree. The list includes synaptic densities that are slightly higher in human brains than in mouse brains, a particular type of glial cell that is larger in the cortex of the former, and myriad genes that have been shown to impact how many neurons are generated in the brain during development.[16] It is unlikely that any single difference can be pinpointed as *the* difference that distinguishes the human species from any other chosen as a reference. Rather, what seems to matter most is the degree to which differences accumulate and modify the result.

Another question of degree, orthogonal to and nonexclusive of the first, regards the size of the brain networks: how many neurons compose them in different species? Could differences in cognitive capabilities across species be traced to differences in the absolute numbers of neurons available for signal processing in each node of the various networks? The underlying assumption here is that neurons are the basic information-processing units of brain networks. Inasmuch as the networks are assembled in similar patterns (which they seem to be), then the greater the number of the neurons that compose a network, the more capacity the network should have to process information—as long as the larger network remains viable and affordable. If a larger brain were a scaled-up version of a smaller brain, built with more neurons that multiply the fundamental loops, like a computer assembled with one, four, or twelve cores, then the more processing units that form a brain, the larger its processing power should be.

Over the past fifteen years, my laboratory has been collecting data systematically on the numbers of neurons that compose the brains of different species, humans included, and can offer new insight that I like to think Darwin would have found illuminating. In quantitative terms, it turns out that the human brain, the seat of the human mind, is made in the image of other primate brains in every way we have checked— except that there's more of it.[17] Larger primate brains have more neurons of about the same average size as smaller primate brains, and humans are no different in this regard (Figure 2.1). Larger primate cerebral cortices have more neurons than smaller primate cortices, and the human cortex, with its 16 billion neurons on average, has as many neurons as could be expected in a generic primate cortex of about 1.2 kilograms. We also have just as many cortical neurons as could be expected for any mammal with our number of neurons in the cerebellum, one of those structures that forms loops with the cortex. Our cerebral cortex is exactly as folded as could be expected of a generic mammalian cortex with its surface area and thickness,[18] and uses the amount of energy—20 percent of all energy consumed by the body—that could be expected for its number of neurons.[19]

Even the size of the human brain is just as expected for the size of our body. Sure, the human fossil record demonstrates that the earliest members of our lineage had brains that were roughly the size of the brain of

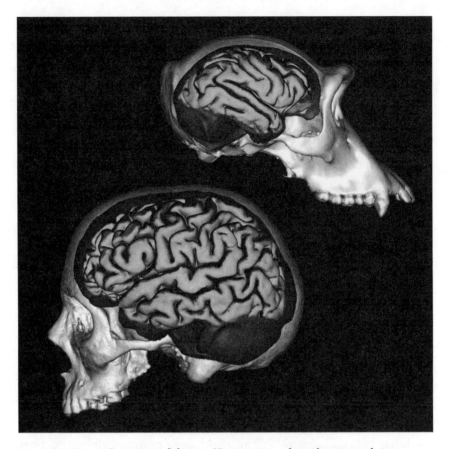

FIGURE 2.1. Brain size and shape in *Homo sapiens* and our closest ape relative, the chimpanzee. (Image courtesy of Aida Gómez-Robles and Jose Manuel de la Cuétara)

a modern chimpanzee; over the past 2 million years, the largest brain of any *Homo* species tripled in volume. Paleoanthropologists have observed that body size also increases over this time, and it is possible that brain increase is, in part, just a by-product of body-size increase. Even enlarged over its evolution, the human brain amounts to about 2 percent of body mass, as in most other primates. Here, it is great apes—gorillas and orangutans—that are the exception. Because larger animals tend to come with larger brains, the fact that gorillas and orangutans can be up to three times the size of a modern human but have brains that amount to only one-third the mass of ours was taken as evidence for decades that the outlier—the special one—was the *human* brain, which is about seven times as large as it should be for our body size. Changing the reference changes the conclusion, however. Compared to most other primates, our brain is just as large as it should be, making the brains of great apes smaller than would be expected, given the trend of primate brains being 1 to 2 percent of body mass. Gorillas and orangutans have retained relatively small brains despite their larger body size. We found that there is good reason for that: these great apes are at the limit of the food-gathering capacity of primates in the wild,[20] given that all of us primates spend eight to nine hours per night lying down and still have other business than eating to tend to during the day, and larger brains would require more energy input in the form of food. So, yes, humans and great apes are dramatically different in the proportion between brain size and body size. But it's not humans who deviate from the primate norm; it is the great apes who were apparently forced to stick to not-very-large brains, retained despite, or exactly because, of their larger body size.

Ancestral humans and ancestral gorillas and orangutans seemingly went different ways in how their evolutionary paths were impacted by the trade-off imposed by a strict limit to how much energy is available through diet to power their brains and bodies. Great apes enjoyed the benefits of investing in larger and larger bodies, while paying the price of not affording a proportionately increased number of neurons; *Homo* stuck to the lean body shape of smaller primates, making more energy available to feed increasing numbers of neurons, especially once simple stone tools and eventually sophisticated cooking technologies were invented. One is not a better path than the other; both were successful in making it to now. Once *Homo* species cooked their food, they overcame the energetic constraints that otherwise apply to larger apes, and the possibility of a larger primate brain opened up to them in a trend that continued until the largest *Homo* brains had nearly tripled in size.

And so, here we are, the primate species with the largest brain and therefore the most cortical neurons among primates. Being a primate, in this case, turns out to be a very important distinction, for the non-primate way of putting brains together involves neurons that quickly become larger as they become more numerous.[21] As a consequence, a primate brain or cortex has more neurons than its non-primate counterpart, and the larger the brain, the larger the difference in favor of the primate species. A gorilla's cerebral cortex is somewhat smaller than a giraffe's but has five times as many neurons. The human cortex has three times as many neurons as an elephant's cortex, even though the human cortex is only half as large. But at about twice the size of a gorilla's cortex, the human cortex also has about twice as many neurons. Humans indeed are more similar to other primate species than any primate is to other mammalian species. We do have a brain made in the image of other primate brains. I like to think that Mr. Darwin would have appreciated that news.

To the best of our current knowledge, the most consequential distinction between human brains and other brains is that we have the largest number of neurons in any cerebral cortex, even including those of the largest whales. That is not because humans stand out from other mammals in how their cortex is built but simply thanks to the double distinction of being the species with the largest brain that is also built the primate way. Just as Darwin expected, there is a gradation in brain size and number of cortical neurons across mammalian species; it follows a relationship that can be described mathematically by a simple power law, and humans fall right along the line.

With more neurons in the cerebral cortex, including its prefrontal associative areas[22] and the hippocampus, it is only to be expected that a result will be an improvement in those cognitive capabilities that depend on representing the past and learning to make predictions from it, forecasting different scenarios, and choosing one according to the best criteria at hand. One of these capabilities is the very working definition of intelligence, according to physicist Alex Wissner-Gross: the ability to choose the course of action that leaves most possibilities open.[23] A new species with more neurons in the hippocampus and connected cortical areas, including the prefrontal cortex, should gain a boost in its decision-making capabilities and thus intelligence, as so defined. Curiosity, which stems from identifying patterns and breaches in them, investigating the sources, finding the problem caused by the changed pattern, and then solving it, should also grow with the number of cortical neurons (although it is doubtful that the same logic applies at the much smaller scale of individual differences *within* a species, where environmental factors and opportunities are known to have huge effects, as examined later, in Chapter 5 of this book). The same logic applies to all other mental faculties that benefit from the complexity and flexibility that cortical processing brings to the table, faculties also long underrated in other animals. If there is a gradation in numbers of cortical neurons across species, there should also be a gradation in the cognitive capabilities they underlie. If one defines intelligence more broadly as behavioral flexibility, as I do, then it is only to be expected that the more the cortical neurons across species, the greater its capability for flexible and complex, and therefore intelligent, behavior.

Indeed, ever since Darwin, nonhuman primates, then non-primate mammals, and more recently birds have been awarded higher cognitive status than even Darwin himself might have suspected. He did consider that similar emotions were expressed across species, and we now know that the circuits underlying anger, fear, joy, and pleasure are very much the same. But shared across humans and other mammals are also maternal care, deceit, self-recognition, planning for the future, playing, learning by imitation, using and making tools, cooperating in problem solving, having a sense of beauty, and even appreciating the taste of food touched by fire. All mammals that have been examined have REM sleep, when cortical representations are reactivated from the inside and humans are known to dream. So it is only reasonable to expect that other mammals also dream, as the twitching and whimpering of sleeping pet dogs appears to confirm. And if they dream, then all that is required for imagination—an ability to conjure images in the absence of their stimulation through the senses—is there. Considered together with the finding that the self-representation network is also there in other animals, it is only to be expected that other mammals also have gradations of self-consciousness. We humans just happen to have loads of it, courtesy of 16 billion cortical neurons organized in loops within the giant loop that is our nervous system.

Even more importantly, we also know by now that the intellectual benefits of having larger numbers of cortical neurons are not strictly limited to biology: they expand into a culture that transcends individual living experiences. Contrary to expectations that larger animals live longer as their metabolic rate decreases (which would supposedly slow down the rate at which damages accumulate), it turns out that maximal life span is extended together with the number of neurons in the cerebral cortex, regardless of body size or metabolic rate.[24] The greater the number of cortical neurons in a warm-blooded species, whether bird or mammal, the longer the maximal longevity of the species. This recent finding explains a number of contradictions in the former metabolic-rate-based framework, such as how a 300-gram cockatoo can live for as long as fifty years, but a lab rat of similar body mass will not make it past three years. It also accounts for why gorillas have a shorter maximal life span than our species, about half the length, even though, with bodies two to three times heavier than ours, their metabolism is slower.

Increased longevity comes along with lengthened childhood, too—another characteristic that used to be considered distinctive about the human species, a derived evolutionary feature. But, again, this may have been because the measuring stick was gorillas and orangutans. If they are larger than us, they should mature later than we do. But it is humans who mature later, at about thirteen years of age, versus eight in those great apes. However, when numbers of cortical neurons are used as the metric to predict age at sexual maturity and maximal longevity, then humans once again fall right along the line, as do great apes. Yet again, humans are not special animals; we are as slow to mature and to age as could be expected of any warm-blooded animal with our number of cortical neurons.

Why more cortical neurons are accompanied by longer lives is a completely new and open question. We first have to ask whether there is any causation involved; it could be that more cortical neurons are just a very good proxy for something else that really matters in extending life. It could also be that damages accumulate stochastically at the same rate across all warm-blooded animals, regardless of their specific metabolic rate (my favorite hypothesis at this point), and so having more neurons—the one type of cell in the body that does not replenish itself during our lifetimes—endows cortical networks with more resilience to accumulating damages. With more cortical neurons still healthy enough to keep body and mind functioning as an integrated whole, life would keep going on much longer.

However it happens, though, the fact that those animals that have more cortical neurons also are slower to mature and live longer has enormous consequences. Without going down the rabbit hole of what comes first or what drives what, it seems fitting that longer-lived animals have more cortical neurons that allow them to deal with the increased amount of information that they will encounter over their extended lifetimes, starting with their longer, parent-sheltered childhood. Likewise, the increasing knowledge accumulated over a longer life gets to be passed down to a next generation that has both more neurons and more time to assimilate it. The story of human evolution thus gains a new dimension with the near tripling in brain size and therefore number of cortical neurons over the past 2 million years. It is a story of presumably increased cognitive capabilities, yes, but also of delayed maturation and increased longevity, and therefore of increasing overlap between generations and, with it, growing opportunities for cultural transfer, learning, and systematization of knowledge.[25] Biology changes, and culture benefits from it.

The implications of accepting the biological and evolutionary continuity between humans and all other mammals are manifold, starting with the fundamental realization that if the same principles that orchestrate behavior in other animals apply to us, with the difference that we have enough neurons to fully represent our own predicaments and use our past to interfere in our future, then there is no longer reason to believe that consciousness is exclusive to humans. Instead, it must differ in degree, not quality, across species, depending on how neuron-rich their cortical circuits are and thus how complex and intricate their representations of self and world can be. I have been asked where to draw the line: What animals have enough cortical neurons that they should be granted personhood and humanlike rights? How many is enough—1 billion? That would include monkeys and all cetaceans but leave out dogs. Or should it be 5 billion, which would separate only elephants, great apes, and the largest-brained cetaceans, besides humans, from the pack? In recognition of gradation, I prefer to argue that the only reasonable course of action is to be respectful toward all creatures, the neuron-rich and the neuron-poor. While it is fine to focus on what distinguishes humans from all other creatures, I prefer to emphasize how biology and neuroscience teach humans to be more humble toward the world, to encourage the appreciation that what has come to distinguish modern humans from all other animals is the culture and technology accumulated over hundreds of thousands of years, and faster and faster at each generation. Thanks to all the achievements those 16 billion cortical neurons in our brain have earned us, human biology is no longer enough to make us modern humans. It takes learning, systematizing the know-what and know-how in past and current generations, figuring out ways to pass it on, and then learning some more. But all the while, we never stop being primates—exactly as Darwin told us.

3

The Darwinian Road to Morality

Brian Hare

Any animal whatever, endowed with well-marked social instincts, would inevitably acquire a moral sense of conscience, as soon as its intellectual powers become as well, or nearly as well developed, as in man.

CHARLES DARWIN, *THE DESCENT OF MAN*, CHAPTER 3[1]

DARWINIAN EVOLUTION DESCRIBES a natural process that accounts for the diversity of all life—including humans. All living things descend from ancestors that leave offspring. These offspring come in a range of varieties, but most are eaten, beaten, starved, or sickened without leaving their own offspring. Only parents pass on their traits to the next generation. These inherited traits can also become modified over successive generations. Some disappear as they no longer convey advantage, and new, better-performing versions are passed on to the next generation. Over millions of years, this same process produced countless species that blanket our planet with life. It continues to shape our world today.

The idea of evolution through natural selection rocked the Western world.[2] In an act akin to moving Earth from the center of the solar system, Darwin replaced an intentional creator(s) with a purposeless natural force. Humans were not the center of creation but equals with all living species. Life did not move toward human perfection but toward the optimization of reproduction.

Although Darwin intentionally avoided discussing humans in *On the Origin of Species*, the implications were apparent. Victorian reaction was polarized. Those who rejected Darwin's premise scoffed at the thought of humans as oversize, hairless monkeys. Those who embraced "survival of the fittest" used it as a powerful metaphor for celebrating everything from individual competitive ability to justification of social stratification.[3]

In *Descent of Man*, and especially in Chapter 3, "Comparison of the Mental Powers of Man and the Lower Animals—continued," Darwin articulates how these views were far too simplistic. He clearly believed concern for others provided a survival advantage, "for those communities, which included the greatest number of the most sympathetic members, would flourish best, and rear the greatest number of offspring."[4] Having overcome his reticence to take on the challenge of explaining the evolution of our own species, in this third chapter Darwin provides the first attempt to account for human morality using evolutionary logic.

His central argument is that human morality is built from different components that were inherited through descent and then modified in our own species by natural selection. He suggests that many of these moral pieces can be seen in the behavior of other animals. The different components he describes include sympathy, regret, reasoning, social learning, love, and self-control. He then uses observations of different cultures to suggest the advantage that morality would confer as well as the unifying features found in all humans. The key premise is that it is friendliness and cooperation that explain some of the biggest evolutionary successes—including that of our own species.[5]

What Apes Share

Cognitive research with a variety of species supports Darwin's views of moral evolution, but our two closest living relatives, the bonobo and the chimpanzee, play a central role. As recently as the 1980s, most scientists believed that bonobos and chimpanzees were distant relatives to our lineage, but we now know they share more of their genomes in common with us than they do with gorillas.[6]

Our two primate first cousins are key to evaluating Darwin's ideas, because they help us identify what traits we share with them through common descent and what features evolved in our lineage since we last shared a common ancestor. When humans, bonobos, and chimpanzees all share a cognitive ability, we know our three species inherited it through common descent. When only humans possess a cognitive trait, we can infer this type of cognition has been modified since our lineages separated. Decades of research have shown, as Darwin predicted, that many cognitive components that allow for our moral behavior are shared through common descent with these close ape cousins.[7]

Ape Sympathy

Darwin proposed that our sympathies are social instincts that come into conflict with one's own needs. They are born from an individual's own previous suffering that is recognized in others. In response, social motivations overcome selfish needs and result in helping. Darwin believed this kind of sympathy was common in animals: "Many animals . . . certainly sympathise with each other's distress or danger."[8]

Sake was four years old when she was released into a room full of delicious fruit at breakfast time. Normally she would just eat the feast of mango, papaya, and banana lying at her feet. But we had designed a situation that would force her to decide between competing motivations. She could fulfill her own selfish desire to eat all the food, or she could open a door for her neighbor and share her meal with another hungry bonobo. Much in the spirit of Darwin's idea about morality originating from situations of conflicting interests, we wanted to see how strong Sake's social motivations truly were. In this case, Sake opened the door for Elike, who ran into the room and joined her, both grunting in happiness as ripe fruit dripped down their overflowing cheeks.

When we tested over a dozen more bonobos, they all showed the same preference, to eat together instead of eating everything alone.

Young and old, males and females opened the door for one another. They did so without ever fighting.

This readiness to eat together made us curious as to whether bonobos have preferences for sharing with some individuals over others. Imagine you were in a casino and you actually won $100. What would you do with this unexpected windfall? Would you find a way to share it with a family member or friend? Or would you use the money to help a complete stranger? Obviously, most people would use the money to do something for or with someone they know, but we wanted to know what bonobos would do if we put them in the same scenario.

We again put Sake in a room adjacent to a room filled with delicious fruit. As she watched us place the fruit, she could also see her groupmate Elike behind one door leading into the food room, but behind another door was a bonobo she had never met. She could choose which bonobo to release and share food with. Sake opened the door for the complete stranger. They hugged, walked over to the food while still embracing, and sat down together to eat. When we tested the other bonobos, they too preferred to share with a complete stranger over their groupmate. It seemed the benefit of making a new friend outweighed the cost of losing some food. Social motivation wins over selfish needs here.

Even stronger evidence for sympathy in bonobos comes from situations in which bonobos help for no benefit. We found that bonobos still helped others when they received no social or food reward. Regardless of whether they were groupmates or strangers, bonobos will proactively remove a barrier between another bonobo and food. They did this even though they were in a different room and could not join the bonobo they helped in eating. In other contexts, bonobos have even been observed to hand their own food over to another bonobo. In some cases, they do this even when the other bonobo does not use gestures or vocalizations to request help. All of this behavior is consistent with Darwin's idea of animals having the ability for kindness in situations where the troubles of others remind them of their own previous suffering. This allows them to overcome their own selfish needs and help others. The sharing among bonobos (and similar behaviors seen in our other closest relative, the chimpanzee) also suggests this component of our moral behavior was already present in our common ancestor.[9]

Ape Reasoning

Advanced reasoning skills are an important part of Darwin's model for the expression of human morality. Inferring what others perceive, intend, and believe allows us to assess the cause of suffering so we can respond. We assign blame if someone fails to help after seeing another person in trouble, but we do not blame them if they never saw the problem. We attribute guilt by inferring the intentions of others. We feel very different about someone unwilling to help than someone who is unable. We can recognize a lie, because we understand we can be led to believe something that is false.[10] Evidence suggests we are not alone in making these types of social inferences that are fundamental to moral judgment.[11]

Kikwit was in trouble, and he knew it. Two adult female bonobos, Semendwa and Bandundu, were frustrated with each other and had begun screaming. Females bonobos almost never direct aggression toward other females. Instead, they look for a motherless male to redirect their anger toward. I watched as they scanned the group and targeted Kikwit. He ran before they even began to chase him. They were faster, but just before they reached him, Kikwit spun to face them. Unlike a male chimpanzee, he did not defend himself. Instead, he looked into the depths of the forest and gave a guttural bark. He followed this alarm call with a threat gesture and slapped the ground with his foot, all while intently staring off toward some unseen enemy. Everyone in the group stopped and joined Kikwit in making alarm calls. Semendwa and Bandundu joined in too. When they all calmed down, Kikwit himself had slipped into the forest. Of course, no threat ever materialized, and the bonobo that cried wolf had escaped unscathed.

Once you have watched a group of primates long enough to know their identities, a soap opera unfolds in front of you. Bonobos and chimpanzees have particularly complicated social lives. It is hard not to attribute to them the same abilities we have in making complex social inferences. Recent cognitive experiments support our intuitions.

Bonobos and chimpanzees carefully monitor what others perceive. If one turns its head suddenly to look at something surprising, other apes will follow, looking in the same direction to see what the first one saw. They will even move around a barrier that is blocking their view of what others can see. When two chimpanzees are competing for food, a more subordinate chimpanzee will approach a piece of food hidden from a dominant's view, instead of food they both can see.

Both species also distinguish between those who are unwilling and those who are unable to help. Chimpanzees become much more frustrated when an experimenter intentionally takes food away than when he or she accidentally drops the food. Bonobos discriminate between an individual who purposely prevents someone from reaching their goal and an individual who tries but fails to help them succeed.

Kikwit's deception was likely intentional. Bonobos and chimpanzees also know when another individual believes something that is not true. For example, they can recognize when someone else thinks food is in its original hiding place because that individual did not see it moved to a new location. They can use this type of understanding when competing. Both species are more likely to steal food or objects from an individual who cannot see them. They also will take indirect path(s) to their goal(s) just to hide their intentions from those who might otherwise try to prevent them from eating or mating.[12]

All of this amounts to solid experimental evidence that both of our close primate cousins have impressive skills for social inference that might have surprised even Darwin. The reasoning needed for basic moral behavior was also likely inherited by common descent with other apes.

Ape Regret

Darwin points to the existence of a conscience in humans as another key ingredient to our moral minds: "Man will then feel dissatisfied with himself, and will resolve with more or less force to act differently for the future. This is conscience; for conscience looks backwards and judges past actions, inducing that kind of dissatisfaction, which if weak we call regret and if severe we call remorse."[13] Are there data suggesting animals (other than cartoon animals) can reflect on past choices and feel regret?

I sat in front of Tai the chimpanzee and waited for her to choose. She knew I had hidden food in one of two places, but she did not see where.

I tried to help her with a clue, by looking in the direction of the correct spot, but even after a dozen or more trials she had not understood my attempt to communicate. She scratched herself nervously and made her choice. I revealed the hiding spot. Tai screamed and slapped her forehead hard, as if in disbelief she had missed again. On the surface it sure looked as if Tai regretted her choice, but it could have easily been an idiosyncratic display of frustration.

Do apes even have the ability to think about past events? Experiments have revealed that bonobos and chimpanzees have detailed memories of the past. In one test, fruit Popsicles and pieces of fruit were hidden in different cups while the apes watched. They had a strong preference for Popsicles, but they were not reminded where they had seen the different foods hidden before they were allowed to pick a cup. If they were allowed to choose after a minute or so, the apes always chose the frozen treat, but if they had to wait an hour or more, they chose the fruit pieces. Bonobos and chimpanzees understood that the frozen treat would have melted over the time they had to wait. This is the type of memory required to reflect on when and where things happened.[14]

Apes may also experience regret. Members of my research team gave bonobos and chimpanzees a choice between a risky and a safe option. They received either a big piece of banana or an unenticing leaf if they chose the risky option, but they always received a few peanuts if they chose the safe option. In each trial, they had to decide whether they wanted to gamble for the big payout or take the acceptable but uninspiring reward. When we analyzed their choices, we found a pattern consistent with regret. In trials where the bonobos and chimpanzees gambled and lost, they quickly tried to switch their choice. They did not try to switch when they chose the safe option to start with or won the big reward gambling. This pattern of choice switching was also strongest in the individuals that were least likely to gamble. Thus it appears bonobos and chimpanzees can show a type of regret, and the individuals least likely to take chances show the most evidence of this emotional response.[15]

The mental abilities required to recall past choices and regret them also appear to have their origin deep in our lineage before we split from the other apes.

Ape Aping

Darwin recognized that human morality is based on a shared set of norms or "codes of conduct" that must be learned from others, writing, "The common opinion how each member ought to act for the public good, would naturally become to a large extent the guide to action."[16] He suspected that animals also might have socially learned behaviors and that evolution eventually shaped these social learning skills into our ability to learn social norms.

The caretakers called the electric fence a psychological barrier. Physically imposing, it was 4.5 meters high and strung with cables that pulsed 40,000 volts of electricity. The chimpanzees escaped at will anyway. I was working at a sanctuary for victims of the illegal bushmeat trade outside of Pointe Noire, Republic of the Congo. All the chimpanzees are orphans, rescued as infants and given lifetime care. That means they have a whole life to learn about the fences that enclose the massive forests in which they live. Much of the learning comes from observing others. They can learn where to safely jump over, squeeze under, or through the fence by watching others. One group of chimpanzees had taken escape to a whole new level. They used green branches to short-circuit the fence. They grounded several small limbs in the dirt and quickly placed them against a single cable. The sticks popped like firecrackers as the water inside conducted electricity into the ground, causing the fence to overload. With the electricity off they easily escaped. Typically, the goal of the escapees was to pilfer the staff's refrigerators or jump into a different enclosure to visit friends.

Darwin knew that great apes could use tools, but he wrote that making tools was one thing that distinguishes humans from other apes. Jane Goodall became famous when she watched as the Gombe chimpanzees of Tanzania proved Darwin wrong. They fashioned grass into termite-fishing rods. Termites trying to defend the colony grabbed onto the grass when the chimpanzees probed their mound. Removing the grass, the chimpanzees had a high protein snack that made up a significant portion of their diet. The discovery led Louis Leakey to declare, "Now we must redefine tool, redefine Man, or accept chimpanzees as humans."[17] Decades of research have revealed that bonobos and chimpanzees make and use dozens of different types of tools—everything from umbrellas for the rain, sponges for drinking, leaf shoes for climbing over thorns, hammers for cracking nuts, and spears for killing small prey. There is also a cultural pattern. Gombe chimpanzees make and use tools more similar to those used by their groupmates than tools used by non-groupmates and vice versa. This pattern mimics cultural traditions in humans that are learned with each new generation and passed to the next.[18]

Scientists today are debating whether it is time to redefine culture or what is human or to accept other apes as cultural. The epicenter of the debate focuses on how traditions are learned. "Aping" is the act of copying another individual. So, it might be surprising to learn that there has been controversy around whether other apes are capable of imitating. Many of their most complex tool-use behaviors might be learned individually or through much simpler forms of learning. It is only in the past decade that so-called "two solution" experiments convinced most skeptics. In these tests, a chimpanzee demonstrates one of two ways to use a tool while her naïve groupmates watch. When given a turn, the observers typically use the tool the way they saw demonstrated. Naïve chimpanzees from a different group who saw their demonstrator use the alternative tool technique copied that method instead. The different groups conformed to the specific technique they saw demonstrated.[19]

This type of social learning and conformity is a critical ingredient to normative behavior that Darwin pointed out as essential to both culture and human morality. Another moral building block seems to have been in place for evolution to play with as our species evolved.

Ape Morality

Darwin suggested the improbable: a purposeless process created morality. He predicted the seeds of our moral code would be found in other animals, positing, "The difference in mind between man and the higher animals, great as it is, is certainly one of degree and not of kind."[20] If this is correct, it would provide key evidence that human morality was built from components inherited through common descent. Today, we have strong evidence for the type of sympathy, reasoning, regret, and social learning Darwin suggested must exist in some simpler form in other animals. Our last common ancestor with the other apes had plenty of cognitive material for natural selection to modify.

Cognitive skills needed for morality did not simply arise completely out of nowhere with the appearance of humans. Yet, no one is claiming that bonobo and chimpanzee societies are guided by anything remotely like a human moral code. Within a sea of similarities between our species, there are real differences. Bonobos have a total lack of sympathy when it comes to sharing objects. They have not been seen giving tools or toys to others in need. Chimpanzees struggle with the simplest types of social inferences needed to cooperate and communicate with others. If you indicate with a gesture where food is hidden or to an object you want them to retrieve, they will not understand your cooperative communicative intention. There is no evidence that bonobos or chimpanzees experience any form of guilt—the type of regret one feels when you fail to help others (as oppose to failing for yourself). This is a basic emotion that is fundamental to many moral interactions. And while bonobos and chimpanzees can imitate, they prefer not to. In most social learning situations, they prefer to generate their own solution to a problem instead of conforming to a demonstrated method. Even if they did have the ability to recognize social norms or codes of conduct, they would be unlikely to follow them.

How then could the moral building blocks we observe in other species have been shaped in our own lineage to produce our unique potential for morality, cooperation, and culture? This is the real challenge Darwin leaves us in Chapter 3 of *Descent of Man*. It is not enough to explain what might allow humans to be social in ways that other animals cannot be. It is also necessary to explain how these social sensibilities appeared in the first place. Other animals, especially dogs (Figure 3.1), are helping us to address this. By studying how natural selection has shaped the cognition of other species we can then make inferences about our own species' cognitive evolution.[21]

FIGURE 3.1. A young orphan bonobo named Masisi, at Lola ya Bonobo Sanctuary
in the Democratic Republic of Congo, being cared for in part by Mystique the dog.
Bonobos at the sanctuary participated in many studies described in the chapter.
Darwin formulated his ideas about moral evolution before bonobos were even described
by Western scientists in the 1930s. I imagine he would have enjoyed learning how much
they have taught us about the evolution of cooperation and morality.
(Photograph courtesy of Vanessa Woods)

Self-Domestication

The evolution of dogs from wolves through domestication gives us a
way for thinking about how natural selection may have shaped our cognition after our lineage split from the other apes.

My childhood dog, Oreo, was my best friend. If I went on a bike ride,
he went with me. If I played in the backyard, he was with me. His favorite activity was to play fetch. He could even fit as many as three tennis
balls in his mouth. He would place them at my feet and bark until I
threw them. When I did, he would instantly run after the first one,
which meant he did not see where the second and third ball were
thrown. When he returned with the first ball, he would bark until I extended an arm in the direction I had thrown the other balls. He then

would run off and orbit in the vicinity of where I had pointed until he could smell the ball. When he brought all the balls back the game would begin again. I never thought anything of this interaction as a kid, other than that it was great fun.

When I reached college, I learned that human infants have a revolution in their social understanding before the first year of life. They begin to understand and use pointing gestures. It is the first way we all begin making social inferences about others. This first type of inference becomes the social foundation for all forms of cultural cognition to follow—including language. Even more surprising to someone who had grown up with Oreo was the thinking at the time that this skill exists only in humans. As sophisticated as bonobo and chimpanzee social reasoning can be, these primates do not understand when you try to help them with your gestures. If you hide food in one of two places and then try to tell them where you hid it by pointing, they can only guess. Even after you do this many dozens of times, they are correct only half the time, or at chance levels.

That I had spent my childhood playing this exact game with my pet dog led my colleagues and me to discover how similar dogs are to human infants. In direct comparisons, only human infants and dogs, but not chimpanzees, are able to spontaneously read the gestures of a human. Like human infants, dogs can read novel gestures they have never seen before and can even understand if you direct them with your voice only. Understanding gestures is also a separate skill from other types of cognition in dogs and human infants that, again, is not present in chimpanzees. Controls in experiments run to test this ability rule out the possibility dogs are just using their strong sense of smell.

How could it be that such a distantly related animal has the ability to make a type of social inference that is a vital developmental marker in our own species? The answer was in Siberia. I traveled there to work with a unique population of foxes that had been experimentally domesticated. The foxes of each generation were selected for breeding based on a single criterion: how friendly they were to humans. Foxes that approached and wanted to be touched by or play with humans were bred together. Those that were fearful or aggressive did not breed. To make sure they could quantify the effects of the breeding program, the researchers also kept a control line. These foxes were bred randomly, without regard for how they interacted with humans.

After only a few generations, the researchers started to see changes in the foxes of the friendly line. They started to approach humans and wag their tails. Some of them had a few white spots on their fur. By generation thirty, the friendly foxes had a high frequency of floppy ears, curly tails, shorter faces, smaller teeth, and multicolored fur coats. None of these changes to their body had been directly selected for in breeding the foxes. Selecting for friendlier foxes had caused changes to appear that are commonly associated with the domestication of dogs from wolves. The fox work shows that domestication is the process of selecting for friendliness.

This meant we could test whether domestication, or an increase in friendliness, gave dogs their unusual ability to understand humans. When I tested the foxes for their ability to respond to human gestures, they performed as well as dogs, even though they were never bred based on their communicative abilities. The control line foxes performed at chance, like chimpanzees. It seems that the ability to cooperatively communicate is linked to selection for friendlier behavior—just like all the other bodily changes caused by selection for friendliness.

Wolves are also more like chimpanzees than dogs in reading human gestures, while even the youngest dog puppies can already understand our gestures. Dogs do not inherit their ability to read humans from their last wolf-like ancestor, and understanding us does not require tremendous exposure or experience with humans. It seems domestication has prepared dogs to interact and understand us in ways other species cannot.

Genomic comparisons between wolves and dogs have revealed that dogs began evolving long before humans planted their first crops. Dogs evolved while all humans were hunter-gatherers. No group of foragers would have purposefully selected a population of wolves to be friendlier for generations, as we saw in the Russian fox experiment. Instead, the fox experiment points to natural selection acting on wolves. Wolves that were able to approach and feed on human waste around settlements would have a huge advantage. They would have access to a very reliable, low-risk source of high-quality food. This set up the exact same selection pressure we saw engineered with the foxes. Over generations, the friendliest wolves would have bred together, since they would be the only ones able to live near humans. Their morphology and social communicative abilities were altered as well, giving humans a way to distinguish them from the other type of wolf, with which they were still competing for food. The wolves that were able to live in proximity to our ancestors chose us. By doing so, they effectively domesticated themselves, or self-domesticated.

Dogs show us how a critical cognitive feature for humans has evolved in a similar way in a very distantly related species. Natural selection acting on friendliness created a new type of social interaction between human and dog. The psychology of dogs was forever changed as they became domesticated.[22] It is now possible to think about how this same process might have played out in our own species' evolution. As Darwin predicted, we can see evidence of how natural selection shapes components of our moral psychology in other species.

Dog Love

Darwin was no stranger to the concern dogs can show. "I have myself seen a dog," he wrote, "who never passed a great friend of his, a cat which lay sick in a basket, without giving her a few licks with his tongue, the surest sign of kind feeling in a dog."[23] He also uses the bond between dog and human, as well as that of parent and baby, to assert that love is a crucial foundation to human morality that is shared with other animals. His insight points to a mechanism that selection can act on when friendliness becomes advantageous: "The feeling of pleasure from society is probably an extension of the parental or filial affections; and this extension may be in chief part attributed to natural selection."[24] I certainly agree it feels like love when you have bonded with a dog, and the latest research suggests the bonding between our two species is a result of changes to systems involved with parent and offspring bonding in both species. As a kid, I would scratch Oreo under the chin, and we would lock gazes. Eventually, he would nuzzle into me for a big hug. Whatever was worrying me would melt from my mind.

Research with animals has uncovered a physiological mechanism that drives social bonding in animals. This mechanism controls a parent's willingness to protect his or her young; a social animal's need to be with, groom, or even snuggle up to another of its own kind; and the price an animal will pay to be together with its mate. If we measure love not by words but by deeds, these are all signs of true love.

Oxytocin is a neurohormone that is created in the brain and becomes active throughout our bodies. It is an ancient hormone present in some form in all vertebrates and has been implicated in everything from bone growth to birth. Its role in social behavior has been best characterized through research comparing different species of voles. One species, the prairie vole, forms very close bonds and mates monogamously, while another, the montane vole, is promiscuous and does not form bonds. Researchers found they could turn on and off these rodents' tendency to bond with one another by regulating oxytocin. Montane voles became faithful to and enamored of their partners when given oxytocin, while prairie voles became more promiscuous when their oxytocin system was blocked. Rodent love can be turned on and off, and oxytocin is the main switch.

The oxytocin system has been altered in dogs as a result of domestication. Dogs and humans are the first species known to have a between-species oxytocin connection. The amount of time that a dog makes eye contact with its owner is related to the amount of oxytocin created in its owner and the level of satisfaction the owner feels in the relationship with his or her best friend. Dogs have evolved to hijack our bonding pathway. But the same interactions also affect our dogs. They too have an increase in oxytocin when they interact with us. When we hug, pet, or even just make eye contact, our dogs experience the same type of oxytocin boost that we do—an effect not observed in wolves. During domestication, the same physiological response that occurs between parent and baby evolved between human and dog. Dogs confirm Darwin's suspicion that love is ancient, evolved, and present in many species. This again points to the possibility that selection for friendliness might have had major impact in our own species' evolution as we evolved to be moral and cultural.[25]

Animals in Command

Darwin highlights the role of "self-command" in helping us weigh two options when we must decide between two conflicting instincts. In the face of conflicts between pursuing some self-interested act or helping others in some way, he wrote, "Man thus prompted, will through long habit acquire such perfect self-command, that his desires and passions will at last instantly yield to his social sympathies, and there will no longer be a struggle between them."[26] This is another area where work with animals reveals how natural selection might shape this type of psychology.

Psychologists have repeatedly found that self-command or inhibitory control is vital to all types of problem solving—so much so that childhood levels of inhibition have been linked to success later in life. In the famous "marshmallow studies," a child was left with a marshmallow and asked not to eat it until an adult returned. The ability of children to listen to the adult and resist their immediate desire to gobble up a sweet treat was later linked to many measures of societal success in these same individuals as adults. Many antisocial behaviors are linked to a lack of inhibitory control, while the most generous acts of kindness often require resisting the need to fulfill one's own selfish desire.

Animals rely on inhibitory control too, and this ubiquity of self-control in other species has allowed scientists to compare the different levels that species demonstrate and to examine the evolution of this quality across the animal family tree. To test our ideas for how self-control evolves, my colleagues and me thought up a kind of marshmallow test for animals. We put a treat inside a clear plastic cylinder that was open at both ends. The trick was, the animal had to resist trying to grab the food through the see-through cylinder and instead had to reach around to the open ends to get the food.

It sounds simple, and some animals could do it without any trouble. But other animals, such as Zoboomafoo, the lemur who starred in the PBS television show *Zoboomafoo*, just kept reaching directly for the food, as if the cylinder were not even there. No matter how many trials we gave him, Zoboomafoo could not resist just grabbing for the food. Eventually he learned to reach around, but he always reached directly for the food first, even though his little fist had bumped into the plastic dozens of times. Being an animal TV star does not require much self-control apparently.

Researchers from all over the world used our cylinder test on thirty-six different species, including different birds, apes, monkeys, dogs, lemurs, and even elephants. When we crunched the numbers, Zoboomafoo was in good company. All the small-brained animals like him struggled with their self-control, while the larger-brained animals easily mastered our animal marshmallow task almost immediately. The bigger the brain, the more self-control a species exhibited. And it was not brain size relative to body size; it was just how big the brain was absolutely. This means evolving a bigger brain will result in more self-control.

Body size is the main determinant of absolute brain size. If natural selection favors a larger body, brains also become larger. This also means larger-bodied animals will have more self-control. We know a lot about how natural selection can favor larger body size. Bigger bodies can be advantageous in colder environments, on islands, and in response to predators. In explaining how humans have unusual self-command that allows for our moral behavior, we need only to explain what favored an increase in body size. The daunting task of explaining the origin of human morality now seems more tractable, thanks to learning how natural selection has shaped the minds of other animals.[27]

The Friendliest Human

Darwin anticipated the power of selection for friendliness in Chapter 3 of *Descent of Man*, when he suggested that communities of humans with the most sympathetic members would flourish more than other groups. My own ideas of how human culture, cooperation, and morality evolved follow this line of reasoning as well.

Selection for hormones like oxytocin leads to more strongly bonded animals. Selection for friendliness in animals leads to increases in cooperative communication. Larger-bodied animals have larger brains and more self-command. We know how natural selection can modify three of the main ingredients for more cooperative and moral behavior. These are the core findings that have led to the proposal that humans are also self-domesticated, like our dogs.[28]

Like Neanderthals and other species of humans, our species inherited an extremely large brain and, with it, tremendous self-command. What differentiated us from other species of humans that came before was selection for friendliness toward a new category of social partner. According to this idea, the spark that led to the evolution of our unusual moral and cultural mind was selection for friendliness that altered our ability to bond with a wider variety of humans. Like bonobos, we also have the ability to show kindness to strangers, but unlike bonobos, we prefer strangers we recognize as sharing our own group identity. We are the kindest human to have ever evolved when we interact with those that share our identity—even those we have never met before. We are also the cruelest species if our group or those that share our identity become threatened. We have evolved to love our group like family. As with dogs, this selection for friendliness has enhanced our abilities to cooperate and communicate. We can behave in highly cooperative and even moral ways toward our group members as a result. We can also morally exclude those humans that threaten our group members. We can then accept harm toward those we do not need to behave morally toward.[29]

While the idea of our species being a product of self-domestication, as with dogs, will continue to be tested, it provides a powerful example of how Darwin's ideas and our findings about animal minds give us ways to examine one of the most complex features of the human mind—our morality.

Conclusion

Reading Chapter 3 of *Descent of Man* 150 years after its initial publication is humbling. The questions Darwin faced seem impossible to address. How could a purposeless mechanism have created the purpose that guides humans in what they ought and ought not to do? How could human morality have evolved from earlier forms if no other animal has anything like a moral sense? If nature is "red in tooth and claw,"[30] how could any sacrifice or selflessness improve survival? Darwin's answers in this chapter predate the description of the gene or the scientific discovery of the bonobo and even the first psychological experiment ever conducted. Yet, amazingly, he anticipated most major elements of modern theories of human sociality, cooperation, and morality.

Descent with modification does have the potential to explain the origin of morality, one of the most unique aspects of being human. Darwin was correct that animals would show evidence for components required for moral behavior. Natural selection had raw material to modify in the cognition of our last common ancestor with other living great apes. Darwin was also correct in his premise that animals could show us how natural selection might modify the cognition inherited through common descent to produce our morality. Natural selection has shaped what Darwin framed as reasoning, love, and self-command in other species. This has allowed us to think about how the same process might have shaped our own species over the last 300,000 years of *Homo sapiens* evolution. We leave with the same radical conclusion that Darwin reached: Religion did not create our morality. What we might recognize as morality, in some nascent form, appeared with the birth of our species. And, more generally, animals show us how doing things that help others, or "survival of the friendliest," can be a real evolutionary win-win. It has been and will continue to be the secret to our species' success.[31]

4

Charles Darwin and the Fossil Evidence for Human Evolution

Yohannes Haile-Selassie

The free use of the arms and hands, partly the cause and partly the result of man's erect position, appears to have led in an indirect manner to other modifications of structure. The early male progenitors of man were, as previously stated, probably furnished with great canine teeth; but as they gradually acquired the habit of using stones, clubs, or other weapons, for fighting with their enemies, they would have used their jaws and teeth less and less. . . . As the various mental faculties were gradually developed, the brain would almost certainly have become larger.

CHARLES DARWIN, *DESCENT OF MAN*, CHAPTER 4[1]

CHARLES DARWIN'S THEORY of evolution by natural selection was one of the most remarkable scientific insights of the nineteenth century. This theory not only became the fundamental tenet of subsequent studies in evolutionary biology, it also served as a metaphorical springboard for how we understand human origins and evolution in the twenty-first century. What we know of our evolutionary history today is built upon most of the theories and mechanisms that Darwin had meticulously described in *On the Origin of Species*[2] and *The Descent of Man*, published in 1859 and 1871, respectively.

Even though Darwin avoided addressing human origins in *Origin of Species*, human evolution constituted the main focus of *Descent of Man*, in which he thoroughly explained his views of the evolutionary processes that led to the origin of humans. Darwin's thoughts in relation to human origins were mostly shaped by his insightful embryological, anatomical, and behavioral observations of various extant animals. These observations led Darwin to argue that humans evolved in the same manner as other species. In Chapter 4 of *Descent*, "On the Manner of Development of Man from Some Lower Form," Darwin starts with evidence for human variation both within and between populations. He establishes that human variation is heritable, and then argues that human populations have the potential for exponential growth. In other words, he establishes that humans are subject to the very same natural "laws" that inevitably lead to evolution by means of natural selection. It is perhaps no wonder, then, that in the second edition of *Descent*,[3] Darwin moves this chapter—one that firmly argues that humans are subject to the same natural forces as other organisms—up to Chapter 2.

Chapters 1–3 of the first edition of *Descent* established ways in which humans were similar to other animals, especially apes. In Chapter 4 of *Descent*, he focused on ways in which we are unique. He identified tool use, reduction in canine size, bipedality (walking on two legs), and increase in brain size as the most important characteristics distinguishing us from other primates. However, Darwin could not test his hypothesis of transitional forms or clearly understand the basic processes of human evolution, largely due to lack of evidence (ancient fossils) during his time. At the time Darwin wrote *Descent*, human fossils had been recovered from caves in Belgium, Gibraltar, and the Neander valley of Germany. It was not clear to Darwin and the rest of the scientific community whether these fossils represented a distinct human species or were just pathological modern humans. We know now that all these fossils belong to Neanderthals.

Less than a decade after Darwin's death in 1882, the evidence that he had desired started surfacing. The discovery of fossils from Trinil, Java, toward the end of the nineteenth century was followed in the 1920s by the discovery of fossils from Zhoukoudian, a cave site in China. These fossils are recognized today as members of *Homo erectus*, an early human species more primitive than *Homo sapiens*. Since these discoveries are from Asia, and not Africa, one could have argued that Darwin was wrong when he predicted in Chapter 6 of *Descent* that "it is somewhat more probable that our early progenitors lived on the African continent than elsewhere."[4] Indeed, a contemporary of Darwin, the German scientist Ernst Haeckel, suggested that humans first evolved in Asia.

However, in 1924, an Australian anatomist and anthropologist, Raymond Dart, found in Taung, South Africa, a fossil skull that came to be known as the "Taung Child."[5] It possessed a small, apelike brain and small canine teeth, but evidence from the hole in the base of the skull suggested that it could have moved on two legs. Dart assigned it to a more primitive form of human ancestor, *Australopithecus africanus*. This was the first specimen that brought the attention of human origins research to Africa, although at first it was met with skepticism by European anthropologists, who rejected Dart's assertions that the Taung Child was on the human lineage. Nevertheless, the Taung finding was just the start of discoveries of early human fossils that could be used to test Darwin's hypothesis of brain enlargement, canine reduction, erect position and bipedality, tool use, and African origins.

Today, we have thousands of fossils of early human ancestors (commonly referred to as hominins), mostly from Africa, spanning the last seven million years of human evolutionary history—near the time human ancestors split from the common ancestor we shared with our closest ape relatives. Today, we have a better understanding of the geologic history and age of the Earth. The neo-Darwinian revolution in evolutionary science (combining genetics and paleontology) since the 1960s and the plethora of fossils discovered from Africa and elsewhere in the past five decades have provided opportunities and approaches to develop new paradigms that help us better understand human origins and evolution. However, even in 2021, researchers of human origins are still trying to answer fundamental questions, including some of the same questions that Darwin himself asked: Why did bipedal apes evolve? What was the body form from which our early hominin ancestors evolved? What are the evolutionary relationships among our earliest ancestors, and which ones among all identified early hominin species are on the lineage that led to our species? Fortunately, there is now enough fossil evidence to test some of Darwin's evolutionary hypotheses, especially in relation to the unique features of humans: large brain, reduced canine size, tool use, and walking on two legs.

Today, we recognize that Darwin was correct about all the characteristics that he identified as unique to humans. However, the currently available fossil evidence indicates that Darwin was probably not correct about the sequence of evolutionary processes that led to what we recognize today as human. Darwin emphasized in Chapter 4 of *Descent* that freeing the hands for tool use was "partly the cause and partly the result of man's erect position." He wrote, "Man alone has become bipedal; and we can, I think, partly see how he has come to assume his erect attitude, which forms one of his most conspicuous differences between him and his nearest allies. Man could not have attained his present dominant position in this world without the use of his hands."[6]

This freeing of the hands as a result of bipedal locomotion, according to Darwin, was followed by reduction of the canine size and ultimately resulted in brain enlargement. Now, 150 years later, we can ask ourselves, is this how we evolved? In our evolution, did our ability to use tools precede or coincide with walking on two feet, as Darwin suggested? Are modern humans a result of gradual transformation of a single lineage? Were there transitional forms (neither quadruped nor fully bipedal like humans) between our ancient quadrupedal forebears and us? These were difficult questions to answer in Darwin's time. However, we now have the fossil evidence representing the past 6–7 million years of our evolutionary history that Darwin did not have access to. This positions us to better address some of the questions stated above. This chapter reviews the available fossil evidence for human evolution and looks back to Darwin's ideas on the evolutionary processes that led to the origin of humans. How close (or far) was Darwin from how we currently understand the pathway by which we became human?

A ⊢ Toros Menalla
  ⌊ Koro Toro

B ⊤ Woranso-Mille
  │ Hadar
  │ Dikika
  │ Middle Awash
  ⌊ Galili

C ⊤ Konso
  │ Omo
  │ Fejej
  │ Koobi Fora
  │ East Turkana
  ⌊ West Turkana

D ⊤ Baringo
  ⌊ Chemeron

E ⊤ Peninj
  │ Olduvai Gorge
  ⌊ Laetoli

F – Uraha

G – Makapansgat

H ⊤ Sterkfontein
  │ Kromdraai
  │ Swartkrans
  │ Gladysvale
  │ Drimolen
  ⌊ Malapa

I – Taung

FIGURE 4.1. Map showing sites of early hominin fossil discoveries in Africa.
(Redrawn by the author)

The Fossil Record

Paleoanthropologists, like me, who study human origins and evolution, and archaeologists, who investigate historic and prehistoric artifacts to understand human culture deep in the past, have been conducting research in various parts of the world for almost a century. The focus of this research became Africa, particularly after the discovery of the Taung fossil by Raymond Dart in 1924. Continued investigations since then, especially in eastern Africa and South Africa, have resulted in the discovery of numerous fossils of early human ancestors and artifacts dating back millions of years ago (Figure 4.1).

Most of the early human ancestor fossils known prior to the 1980s were recovered by projects led by non-African scientists, largely because there were no properly trained African paleoanthropologists until the late 1980s. My personal journey to fossil hunting started about that time. I was working for the Ethiopian Ministry of Culture, with a first degree in history, and accompanied a paleontological project to the field as an antiquities officer. That was when I had my first glimpse of a fossil on the ground, even though I had seen a lot of them in the laboratory at the National Museum of Ethiopia. Apparently, I turned out to be a good fossil hunter and was encouraged to pursue my higher education in paleontology. After I received my doctoral degree from the University of California at Berkeley, with one of the most preeminent paleoanthropologists, Tim White, as my adviser, I brought together an international multidisciplinary team of scientists and established the Woranso-Mille paleontological project in 2004 to conduct fieldwork and collect early human ancestor fossils in the Afar desert of Ethiopia.

The Woranso-Mille study area is one of many paleontological sites located in the central Afar region of Ethiopia.[7] We have to drive about 560 kilometers northeast from the Ethiopian capital, Addis Ababa, on a nicely paved road, before turning off-road to get to the site. Most of this region is desolate, dry, and extremely hot, with temperatures rising as high as 120 degrees Fahrenheit (49° C). The local Afar people have been living on this land for millennia, moving with their herds from place to place looking for water and grazing.

The Afar land is far from any developed city; my colleagues and I live in tents for almost two months each year. I am so focused on the fossils that I sometimes forget about the extreme, dangerous heat and continue working throughout the hottest hours of the day. The fossil remains of our early ancestors keep us going. I always have a unique feeling while walking on sediments that are millions of years old, erosion exposing their best-kept secrets through tiny fragments of glittering teeth, jaws, and sometimes long bones. As a paleoanthropologist of African origin (of course, we are all Africans!), it gives me pride and honor to actually find ancient fossil remains of our ancestors in the continent that Darwin predicted they would be found. I have discovered and named a few hominin species in the past two decades, and recent discoveries from the site also promise to elucidate some of the outstanding questions in early human evolutionary history.

Today, more than twenty early human ancestor species have been identified in the fossil record, some of them dated as old as 6–7 million years. However, most of them are known from fragmentary jaws and teeth, and more complete specimens are extremely rare. The most important question for us paleoanthropologists is: How do we identify a fossil as a member of our lineage (hominin)? It turns out that the key anatomies that precede all others in a hominin are a small, blunt (what specialists would call non-honing) canine and evidence for bipedalism.

Paleoanthropologists use the term "hominin" to refer to the group that includes all extinct and living human relatives after the split from the common ancestor they shared with the chimpanzee lineage (see Chapter 6 of this book). Inclusion of any fossil species into this group is mainly determined by the presence or absence of the two features described above. The earliest recognized hominins show these features to some degree, even though they also retain a number of apelike features. In the following sections, fossil hominin species that have been recognized thus far are presented in three groups: 1. the earliest hominins (older than 4.2 million years), 2. *Australopithecus* (4.2–1.0 million years ago), and 3. *Homo* (2.5 million years ago to the present).

The Earliest Hominins (7.0–4.2 Million Years Ago)

In 2002, fossils from 6–7 million years ago were discovered at a place called Toros-Menalla, located in the Djurab Desert of northern Chad (see Figure 4.1 for this and other hominin fossil locations). These fossils, named *Sahelanthropus tchadensis*, belong to the oldest human ancestor known in the fossil record.[8] This discovery in central Africa was a surprise, because almost all early human ancestor fossils have been found in eastern Africa and some South African cave sites. The designation of this species was based on a slightly deformed but complete cranium, jaw fragments, and isolated teeth. It is inferred to have at least occasionally walked on two legs, because the hole at the bottom of the skull, the foramen magnum, is in a humanlike position and not located in the back of the head, as it is in animals that walk on all fours. Unlike apes, it also had reduced, blunt canines. However, it had a very small brain volume (360–70 cubic centimeters), similar to that of living apes such as chimpanzees.

Thousands of kilometers east of Chad, early hominin fossils were recovered from the Tugen Hills in the Baringo district of Kenya in the late 1990s and assigned to a different species, *Orrorin tugenensis*.[9] These fossils have been dated to 5.6–6.0 million years ago. Some of the most important fossils assigned to this species include an upper canine that appeared to be more primitive than a human tooth but also was not apelike. Elements of the lower limbs (femora) of this species were also recovered, and they were interpreted as similar to *Australopithecus* in some features, but more humanlike than *Australopithecus*, probably indicating more efficient bipedality in *Orrorin tugenensis*. However, later research demonstrated that *Orrorin tugenensis* was not more humanlike than *Australopithecus* in terms of its bipedal adaptation.

One of the most important paleontological sites to have yielded significant fossil remains of early human ancestors is the Middle Awash, located in the Afar region of Ethiopia. This site documents a continuous record of our evolutionary history spanning the past 6 million years. In the 1990s, I discovered a jaw fragment with teeth, some parts of an arm, a toe bone, and isolated teeth from 5.2–5.8-million-year-old sediments and assigned them to a new species of early hominin, which I named *Ardipithecus kadabba*.[10] Bipedality in *A. kadabba* was inferred from a single toe bone, which is shaped in a way that would have allowed *Ardipithecus* to push off the ground when it walked on two legs. The canines were as primitive as those of *Orrorin tugenensis* and not apelike. Thus, Darwin's prediction that the earliest members of our lineage were "probably furnished with great canine teeth" is inconsistent with the fossil evidence.

In 1994, fragmentary fossil discoveries from the Middle Awash site led to the naming of a new, 4.4-million-year-old hominin, *Ardipithecus ramidus*.[11] This species is considered to be a descendant of *A. kadabba*, which was found in close geographic proximity but in much older deposits. Most of the skeletal anatomy of the species was unknown until the discovery, in 1994 at Aramis (Middle Awash) of a partial skeleton, nicknamed "Ardi." Detailed analyses of Ardi showed that A. ramidus had a skeletal morphology with a unique combination of traits that paleo-anthropologists had not expected to see in our evolutionary pathway and that challenged the notion that humans descended from a chimpanzee-like ape.[12] Indeed, Darwin predicted such unique ances-tors when he wrote, "But we must not fall into the error of supposing that the early progenitor of the whole Simian stock, including man, was identical with, or even closely resembled, any existing ape or monkey."[13]

Most surprisingly, A. ramidus was a bipedal hominin with an oppos-able big toe! Of course, it is not difficult to realize that bipedality in A. ramidus was not humanlike, since its big toe was used for grasping rather than walking. This species was, rather, a facultative biped that walked on two legs when on the ground but in a unique way. Its discoverers interpreted the locomotor adaptation of the species as "less adept in the trees than are living chimpanzees, but . . . [a] more capable climber and clamberer than Australopithecus."[14] Its hand bones had long fingers but did not show any sign of knuckle-walking, as seen in modern chimpan-zees and gorillas. Its hip bones were humanlike in some respects—showing features that are uniquely associated with bipedality—but apelike in others. The canines of A. ramidus were relatively small, with minimal difference in size between males and females.

Our knowledge and understanding of the anatomy of the earliest human ancestors prior to the discovery of the four earliest hominin taxa described above was based on fossils younger than 3.6 million years. The discovery of these earliest human ancestors since the early 1990s has clarified the earliest phases of our evolutionary history. These species are still among some of the most poorly known hominins in the fossil record, and their anatomy is not well understood yet. However, what has been recovered for each taxon thus far indicates that these earliest hominins show some degree of bipedality (different from that of mod-ern humans), had relatively reduced and non-honing canines, and all had small brains. Furthermore, the available fossil record does not show any sign of tool use associated with any of these species, nor do the known hand bones show entirely humanlike morphology to allow a precision grip. Thus, the story emerging from the fossil record differs from that envisioned by Darwin. It is one in which some form of upright walking and canine reduction preceded stone tool use and brain enlargement by millions of years.

Precisely why our earliest ancestors had reduced canines and began walking on two legs remains unknown. Darwin supposed in *Descent* that our bipedal ancestors "would thus have been better able to defend themselves with stones or clubs, to attack their prey, or otherwise to obtain food."[15] While some scientists still prescribe to some form of this idea, others have proposed that bipedalism was favored for reasons having to do with gathering and carrying food or babies, for energetic savings, or for display. The truth is, we still don't know why bipedalism evolved—it remains a mystery that we hope will be solved as we continue to discover the fossilized remains of our earliest ancestors.

Australopithecus (4.2–1.9 Million Years Ago)

By 4.2 million years ago, a new form of human ancestor appears in the fossil record that is markedly different from the earliest human ancestors mentioned above. This was the beginning of the genus known as *Australopithecus*, mentioned earlier in the context of Raymond Dart's 1924 discovery of the Taung Child. Dart assigned Taung to the species *Australopithecus africanus* and argued that it represented the earliest apelike human ancestor. Later, more fossil discoveries from eastern Africa and South Africa were assigned to numerous species within the genus *Australopithecus*. Compared with the earliest human ancestors, discussed above, *Australopithecus* had much larger cheek teeth (postcanine megadontia) and a broader face, which is associated with even smaller canines. It was also better adapted for walking on two legs than its predecessors, even though its brain size remained similarly small. It is this genus that most researchers interpret as the ancestor of our genus, *Homo*, and other related groups that went extinct at various intervals in our evolutionary history.

FIGURE 4.2. A 3.8-million-year-old cranium of *Australopithecus anamensis* from
Woranso-Mille, Ethiopia, with facial reconstruction. (Photographs by Dale Omori [left]
and Matt Crow [right]. Facial reconstruction by John Gurche.
Courtesy of the Cleveland Museum of Natural History)

The earliest *Australopithecus* fossils, assigned to the species *Australo-pithecus anamensis*, have been known from 4.2–3.9-million-year-old sites in the western Turkana Basin of Kenya[16] and the Middle Awash of Ethio-pia (see Figure 4.1 for locations).[17] My team, however, recently found a nearly complete 3.8-million-year-old cranium of this species from Woranso-Mille, in the Afar region of Ethiopia, which not only provided the first glimpse of its facial and cranial anatomy (Figure 4.2) but also demonstrated that it overlapped in time with its putative descendant, *Aus-tralopithecus afarensis*, for about 100,000 years.[18] Most of what we under-stand about *Australopithecus* in eastern Africa comes from our knowledge of *Australopithecus afarensis*—Lucy's species. *A. afarensis* is one of the best-known, longest-lived (3.9–2.9 million years ago), and extensively studied early human ancestors ever found in the fossil record. It is a species that roamed the East African Rift System, from Ethiopia (Woranso-Mille) in the north to Tanzania (Laetoli) in the south, across a stretch of more than 2,000 kilometers. A 3.5-million-year-old jaw found in central Chad suggests that A. *afarensis*, or something like it, also existed 2,500 kilometers to the west, across north-central Africa. Best known from Hadar (3.4–2.9 million years ago), in the Afar region of Ethiopia, and Laetoli (3.7–3.5 million years ago), in Tanzania—but also known from other eastern African sites—it is represented by hundreds of fossil specimens representing almost every part of the skeleton. Laetoli also documents the earliest truly bipedal hominin footprints, from 3.6 million years ago.

The anatomy of A. *afarensis* is best known from a 3.2-million-year-old partial skeleton nicknamed "Lucy"[19] and a collection of fossils representing about seventeen individuals, nicknamed "The First Family,"[20] both found at Hadar, Ethiopia, in 1974 and 1976, respectively. The 1970s were indeed the golden years of paleoanthropology, and the discoveries from Ethiopia during that decade substantially redirected the course of human evolutionary studies. I was in high school in Ethiopia when these significant fossils were being unearthed from the Afar desert. It was also a time when Ethiopia's long-ruling monarch (Emperor Haile Selassie I)* was toppled by a military junta, and few Ethiopians heard about these fossil discoveries. I became aware of the significance of these fossils only when I went to college and took an introductory archaeology course called "Early Man" during my junior year as a history major—which was really what triggered my interest in human evolutionary studies.

Since the naming of *Australopithecus afarensis* in 1978, based on the fossils from Hadar (Ethiopia) and Laetoli (Tanzania), a significant number of specimens of this species have been found at other sites in Ethiopia and in Kenya. Among the most important newly discovered fossil specimens of A. *afarensis* are a 3.3-million-year-old partial skeleton of a juvenile from Dikika[21] and a 3.6-million-year-old partial skeleton of an adult individual from Woranso-Mille,[22] both located in the Afar region of Ethiopia. As a result of the number of fossils known for this

---

* I am not related to Emperor Haile-Selassie I, who ruled Ethiopia for forty-three years.

species, its cranial, dental, and postcranial anatomy and its locomotor adaptation are better documented than those of any other human ancestor in the fossil record.

Australopithecus afarensis individuals had larger cheek teeth and smaller canines than their predecessors, such as *Ardipithecus* and *Sahelanthropus*, which generally had smaller cheek teeth and larger canines, and these characteristics signify a major change in the dietary adaptation of our ancestors. As the African forests receded and grasslands spread, our ancestors ate less ripe fruit and tougher, more fibrous food. It was also at this time that *Australopithecus* mastered upright walking, as its members had to move across large areas looking for scarce resources. As our ancestors became more bipedal, Darwin predicted in *Descent*, "The pelvis would have to be broadened, the spine peculiarly curved, and the head fixed in an altered position and all these changes have been attained by man."[23] These changes that Darwin meticulously described can be found in the 3.2-million-year-old Lucy and other fossils of her species, *Australopithecus afarensis*. In addition to the shape of the hip bones and lower limbs of the species, the Laetoli footprints clearly indicate that *A. afarensis* was a committed terrestrial biped as early as 3.6 million years ago. However, it may have retained some features that are usually seen in arboreal primates. For example, *A. afarensis* had slightly longer arms relative to its legs—humans have longer legs relative to their arms—and curved fingers and toes that would have allowed it to climb trees more efficiently than we do. However, some researchers have suggested, as an alternative explanation, that the longer arms and curved fingers and toes could also be retentions from a primitive ancestor and may have had no functional significance. The facial morphology of *A. afarensis* was more like that of apes than humans, and it still had a relatively small brain, 450 cubic centimeters. The average brain volume in modern humans is around 1,300 cubic centimeters.

Interestingly, recent discoveries have revealed that Lucy and her kind were not the only hominins on the landscape 3.9–2.9 million years ago. In the past fifteen years, my colleagues and I have found many early hominin fossils at Woranso-Mille dated to 3.0–4.0 million years ago. In 2015, we assigned fossil remains of upper and lower jaws recovered from

3.3–3.5-million-year-old deposits at the site to a new species, *Australopithecus deyiremeda*. This species was distinguished from *Australopithecus afarensis* mainly by the robusticity of its lower jaw relative to the size of its smaller teeth, thicker enamel on its teeth, and less protruding face.[24] Some researchers argue that the difference between *A. deyiremeda* and *A. afarensis* is so minimal that they are really just the same species. We'll see, as more fossils are discovered. Regardless of the validity of *A. deyiremeda*, however, there is now stronger evidence from the same region that a second, if not third, hominin species coexisted with *A. afarensis* in the Afar region. This yet unnamed species is incontrovertibly different from *A. afarensis*—it had a different bipedal locomotor adaptation.

In 2009, my team discovered a hominin fossil known as the "Burtele Foot" from one of the 3.3–3.5-million-year-old localities of Woranso-Mille.[25] This partial foot comprises eight toe bones, including bones of the big toe, which clearly indicate that the foot belonged to a hominin species with a grasping big toe. The contemporaneous *Australopithecus afarensis* did not have an opposable big toe—we know that from the preserved foot fossils of the species and from the Laetoli footprints.[26] What the Burtele Foot shows is that there was a hominin species at the same time as Lucy's species equipped with better ability to climb trees but less fit for walking on two legs when on the ground. The species to which the Burtele Foot belongs must have had an adaptation similar to *Ardipithecus ramidus*, a species more than a million years older. While the taxonomic affinity of the Burtele Foot remains unknown until the discovery of more fossils from Burtele, it is compelling evidence for the presence of more than one early human ancestor species between 3.5 and 3.3 million years ago.[27]

Another fossil hints at diversity in the human fossil record at this time and reveals the difficulty of deciding whether a fossil represents a new species or is just evidence of variation within a species. In 2001, paleontologists working in the Turkana Basin of Kenya named a new species, *Kenyanthropus platyops*, based on a distorted cranium, fragment of an upper jaw, and a small fragment from the ear region.[28] The discoverers claimed that *K. platyops*, recovered from a 3.3-million-year-old locality known as Lomekwi, had a flat face and small teeth, more like members of our genus, *Homo*. In fact, they claimed that this species could be the direct ancestor of *Homo*, which would bump Lucy from her place in the human family tree. Like every other newly named hominin species, *K. platyops* has been under scrutiny by other researchers. Some still accept *K. platyops* as a valid species, but many others have questioned its validity, claiming that the skull of this species, if it were not distorted, would have been like those of *Australopithecus afarensis*.

As Darwin wrote in Chapter 7 of *Descent*, "Every naturalist who has had the misfortune to undertake the description of a group of highly varying organisms, has encountered cases (I speak after experience) precisely like that of man; and if of a cautious disposition, he will end by uniting all the forms which graduate into each other, under a single species; for he will say to himself that he has no right to give names to objects which he cannot define."[29] This is not to say that *Kenyanthropus* has not been defined but only to illustrate the difficulty of determining whether a fossil represents a new species or is just evidence of variation within a species, and to underscore the importance of new fossils and additional analyses in resolving ongoing debates in our field. Generally speaking, it appears that during the time between 3.5 and 3.0 million years ago, hominin populations were experimenting with various habitats and diversifying as African forests were receding and more open grasslands were spreading. This created opportunities for speciation and extinction and could explain the apparent diversity we see during that time.

The fossil evidence clearly shows that hominins dated to between 4.2 and 2.9 million years ago were committed terrestrial bipeds, or spent a significant amount of time on the ground, with the exception of the species represented by the Burtele Foot. We also see that these hominins had much larger cheek teeth and their canines reduced, compared to those of their predecessors. This is a major adaptive shift from the earlier species *Ardipithecus ramidus*, in terms of both their dietary and their locomotor adaptations. At the same time, however, their brain volume remained small, like that of their predecessors. Despite all these changes, however, there is no compelling evidence in the fossil record for frequent tool making or tool use by any of these species.

Recently, some bone fragments, apparently showing "cutmarks induced by hominin activity," were recovered from Dikika, an *Australopithecus afarensis*–bearing site in the Afar region of Ethiopia. The discoverers argue that since *A. afarensis* has been recovered from Dikika, the cut-marked bones might indicate that *A. afarensis* may have used tools.[30] However, this inference is not unanimously accepted and remains controversial, mainly because it has been shown that other animals, such as crocodiles, could leave marks on bones similar to the ones seen on the Dikika fragments.

Another possible, but also equally controversial, discovery of tool use by early hominins comes from stone tools from Lomekwi, a 3.3-million-year-old locality in the Turkana Basin of Kenya, close to where *Kenyanthropus platyops* remains were recovered—and some have suggested that this species might be linked to the tools.[31] Others firmly object, stating that the tools could have been eroded from much younger sediments and reburied in older sediments. These claims that the evidence from both Dikika and Lomekwi represents very early tool use are disputed and remain to be confirmed by more conclusive evidence. We'll see, as more research is done at these and other sites of similar age throughout Africa.

From 3.0 to 1.0 million years ago, *Australopithecus* continued to evolve and to diversify. Human evolution in eastern Africa between 3.0 and 2.5 million years ago is almost entirely unknown, largely due to the lack of fossils from this time period. However, this is an extremely critical time period for understanding the origin of our genus, *Homo*. Thus far, only a single, 2.8-million-year-old jaw fragment has been recovered from Ethiopia and assigned to the genus *Homo* (more on this later). Fossil hominin species found at sites in both eastern Africa and South Africa from 2.5 to 2.0 million years ago are claimed to be the ancestor of the genus *Homo*. However, none of the currently available fossil evidence is convincing, and more fossils need to be discovered from this time period.

In 1997, I found a 2.5-million-year-old cranium in Middle Awash, Afar region of Ethiopia, at a locality known as Bouri, and two years later it was announced as a new species, *Australopithecus garhi*.[32] *A. garhi* resembles the earlier *A. afarensis* in its small brain volume and thick-enameled, large cheek teeth, while it also has some jaw features that are like jaw features of our genus, *Homo*. As a result, the researchers who studied the specimen have claimed that *A. garhi* is at the right time and place to make it the best candidate to be the ancestor of the genus *Homo*. They further corroborate their claim with the possibility of tool use in *A. garhi*, for a few bones with apparent cut marks were recovered from the site where the cranium was found. However, no stone tools were found. On the other hand, the earliest, uncontested stone tools, from 2.6 million years ago, were recovered from a place called Gona, about eighty-five kilometers north of Bouri.[33] The assumption is that the contemporaneity and close geographic proximity of *A. garhi* and the earliest stone tools make it more likely that the stone tools at Gona were made by a species similar to *A. garhi*. Nevertheless, *A. garhi* remains a poorly known species, and its relationship with our own lineage is not fully understood.

Meanwhile, in caves in South Africa, researchers have been discovering fossils ever since Raymond Dart's discovery of the Taung Child in 1924. Discoveries from other South African cave sites such as Sterkfontein and Makapansgat continue to be made, and these fossils have been assigned to the species *Australopithecus africanus*. Most of the specimens assigned to this species are younger than 3.0 million years. However, recent analysis indicates that some specimens, such as one commonly known as "Little Foot" (StW 573), might have been as old as 3.6 million years. The discoverers also think that this specimen may not belong to *A. africanus* and have assigned it instead to a species called *Australopithecus prometheus*.[34] Regardless of the species' name, however, this old age for *A. africanus* and/or Little Foot is contested. Furthermore, in 2008, at a 2.0-million-year-old cave site called Malapa, researchers discovered two partial skeletons (MH1 and MH2) of yet another hominin, which they called *Australopithecus sediba*, and claimed that it is the potential ancestor of the genus *Homo*.[35] This is a less likely scenario, especially since fossils of *Homo* that are much older than *A. sediba* have already been discovered from various sites in eastern Africa. Furthermore, more recent fossil discoveries from a South African cave site known as Drimolen showed that *Homo erectus*, *A. sediba*, and *Paranthropus robustus* (see below) were contemporaneous between 1.9 and 2 million years ago.[36] These discoveries likely exclude A. sediba from the ancestry of the genus Homo. Once again, however, the relationships between these fossils and our own genus, Homo, remain unclear.

This is also a time period when we clearly see the first appearance of a second hominin lineage that was not directly related to the lineage that eventually led to us. Fossils recovered from 2.0–3.0-million-year-old sites indicate that early human ancestors radiated into two major forms: one characterized by large teeth and heavy chewing musculature (commonly recognized as the "robust" lineage), and another with smaller teeth, which may have eventually led to our own genus, Homo. The "robust" lineage (sometimes given its own genus, Paranthropus) survived for at least 1.5 million years and went extinct around 1.0 million years ago.[37] Thus far, paleoanthropologists have identified three different species in this lineage: Paranthropus aethiopicus, Paranthropus boisei, and Paranthropus robustus, the former two found in eastern Africa and the lattermost in South Africa. However, researchers do not think that any of these species could have given rise to ours, because they were all specialized in their dietary adaptation, unlike the hypothesized putative ancestors of the genus Homo, who were more likely food generalists utilizing a variety of resources. Exactly when our genus first appeared in the fossil record, or which earlier Australopithecus species gave rise to it, currently remains unknown.

Homo (since 2.8 Million Years Ago)

We humans belong to the genus Homo. Our genus is distinguished from its earlier ancestors by a suite of cultural and biological characteristics, including obligate bipedality (the only way to move about efficiently is standing erect on two feet), tool making, an enlarged brain, and eventually the use of language. The earliest fossil evidence for our genus comes from localities in Ethiopia dated to 2.8 and 2.3 million years ago.[38] But these are quite fragmentary fossils, and the oldest, more complete remains of the genus are from ~1.8 million years ago: Homo habilis, found at Olduvai Gorge, Tanzania,[39] and Homo rudolfensis, discovered at

Koobi Fora, Kenya.[40] The earliest members of our genus have larger brains and smaller cheek teeth than *Australopithecus*, and they are more often found in association with stone tools, indicating a greater reliance on technology for survival. However, it is unlikely that they had language in the sense that we understand it today.

Soon after the evolution of our genus, *Homo erectus*, a species larger-bodied and larger-brained than *H. habilis* and *H. rudolfensis* appears about 2 million years ago. While the oldest *H. erectus* fossils appear to be from eastern and southern Africa,[41] this species extended its range outside Africa relatively quickly. The 1.77-million-year-old site of Dmanisi, Georgia, has yielded five well-preserved skulls and partial skeletons representing the earliest hominins known to inhabit Eurasia.[42] In addition to increasing its home range, *H. erectus* invented more sophisticated stone tool technology in the form of the hand ax, a tool-making tradition archaeologists call the Acheulean industry. More importantly, perhaps, sites in Kenya and South Africa preserve 1.0–1.5-million-year-old sediments that contain the earliest evidence for controlled fire. Darwin wrote of the ancestral human in *Descent*: "He has discovered the art of making fire, by which hard and stringy roots can be rendered digestible, and poisonous roots or herbs innocuous. This last discovery, probably the greatest, excepting language, ever made by humans, dates from before the dawn of history."[43] Darwin would probably be thrilled and fascinated to learn that fire was tamed by a predecessor of humans, *Homo erectus*.

Throughout the Pleistocene, members of the genus *Homo* expanded their range throughout Eurasia and began to diversify into distinct populations. Even in Darwin's time, fossils were known of a population of ancient Europeans named by the British geologist William King in 1863 as a distinct species, *Homo neanderthalensis*.[44] In just the last two decades, researchers have discovered and named *Homo naledi* in South Africa and the small-bodied *Homo floresiensis* and *Homo luzonensis*, both discovered in caves on islands in southeast Asia.[45] These are species or populations of humans who lived only in the later part of the Pleistocene. However, their place in human evolution in general and their relationship to modern humans remain unknown.

Technological advances and paradigm shifts in the last few decades have changed how researchers interpret our past based on the fossil evidence—they can now extract and analyze ancient DNA from some Pleistocene fossils. Fragmentary fossils found in Denisova Cave in Siberia, for instance, contain DNA that does not match any known hominin and represents a previously unknown population of ancient human called the Denisovans.[46] This revelation wouldn't have been possible just from the visible morphology of the bones.

DNA has also predicted both where and when the earliest members of our own species, *Homo sapiens*, would have evolved: back home in Africa, ~300,000 years ago. Sure enough, fossils from South Africa, Morocco, and Ethiopia are the oldest known from our species, and they date to between ~160,000 and ~300,000 years old.[47] Eventually, our species grew in population and expanded its range beyond the borders of the African continent. Thereafter, we encountered Neanderthals in Europe and Denisovans in Asia. Ancient DNA reveals that we absorbed those populations into our own, meaning that there was a small, but meaningful, amount of interbreeding between these populations—enough that Neanderthal and Denisovan DNA is still detectable in the human gene pool today.

Conclusion

The human lineage has been evolving for the past 6–7 million years, after the split from the common ancestor with chimpanzees. When Charles Darwin wrote *Descent of Man*, he realized that our progenitors diverged somewhere from the catarrhine (Old World monkey and ape) stock. However, his major inquiry was where, when, and how humans evolved. He hypothesized that the last common ancestor of apes and humans lived in tropical Africa, an idea largely influenced by Thomas Huxley's *Evidence as to Man's Place in Nature* (1863). Huxley's monograph detailed how the "anthropomorphous" apes (gorillas, chimpanzees, orangutans, gibbons), and especially the African apes (gorillas and chimpanzees), are more closely related to humans than other animals.[48] Unfortunately, Darwin did not have any fossils with which to test his hypothesis of African origins. The only notable fossils known at that time were from Neanderthals. Fossils that could have been relevant to test his hypothesis of African origins were found only after his death. However, in Chapter 6 of *Descent*, Darwin wrote, "In each great region of the world the living mammals are closely related to the extinct species of the same region. It is therefore probable that Africa was formerly inhabited by extinct apes closely allied to the gorilla and chimpanzee and as these two species are now man's nearest allies, it is somewhat more probable that our early progenitors lived on the African continent than elsewhere."[49]

Today, we have enough fossil evidence to inform us that Darwin was right when he said the birthplace of the human lineage is Africa. Not only that, the current evidence indicates that the origin of our genus, *Homo*, and our species, *Homo sapiens*, also took place on the African continent.

We also have genetic evidence for the time of the split from the last common ancestor we shared with chimpanzees, 6–7 million years ago. However, Darwin thought that the major reason the human lineage became bipedal was to free the hands for tool use, and that this was followed by reduction of the size of our canines and enlargement of our brain. The fossil evidence that we currently have in hand indicates that in fact walking on two legs preceded tool use, canine reduction, and brain enlargement. The reason our lineage became bipedal is still debated, even though a number of hypotheses exist. The earliest hominins underwent a major adaptive shift, where walking on two legs and freeing the hands was advantageous and persisted. Our understanding of human origins continues to improve as more fossils are recovered from Africa and elsewhere. However, we are by no means anywhere near painting the full picture of our origins and evolution, which Darwin and his predecessors inquired about almost two centuries ago, without any fossil record.

5

A Century of Civilization, Intelligence, and (White) Nationalism

Kristina Killgrove

The subjects to be discussed in this chapter are of the highest interest, but are treated by me in a most imperfect and fragmentary manner.

CHARLES DARWIN, *THE DESCENT OF MAN*, CHAPTER 5[1]

IN SEPARATING his groundbreaking idea of natural selection into two different works—*On the Origin of Species* (1859) and *The Descent of Man* (1871)—Charles Darwin mitigated complaints he knew would be leveled at his scientific theory. Treating humans as distinct from other animals allowed Darwin to lay out his careful argument and generate discussion with *Origin*, while waiting for the right time—a decade later—to explain in *Descent* how those processes shaped modern humans.

This separation of humans from other animals has a very long history, with the idea of a Great Chain of Being, dating back to ancient Greek philosophers but more prominently taken up by nineteenth-century European thinkers, to explain the development and progress of life on

Earth. Although Darwin was convinced that evolutionary processes were the same for the entire animal kingdom, he wrote two treatments of evolution to explain animals and humans, in keeping with the scientific tenor of the day. His was a time when anthropology—the current field of research that tackles all questions related to present and past humans, including our origin and especially our culture—had only just begun to exist. In the mid-nineteenth century, the allied field of ethnology was the study of living, non-Western humans, primarily the peoples of Africa and the Americas, from which European explorers were bringing waves of information for scientists to comb through and create natural histories about. A break between British ethnologists and anthropologists culminated in the formation of the Anthropological Society of London in 1863, between the publication dates of Darwin's two major volumes on evolution.

Anthropology, particularly in Darwin's native England, was therefore a latecomer to the research stage, even compared to the young discipline of evolutionary biology. The new field of anthropology encompassed ethnology, and to it were added prehistoric archaeology and craniology, or the study of the size and shape of skulls and their contents.[2] Theology also held sway, with the Bible forming the fundamental source through which most people understood their place in the world. Yet, Scripture was interpreted in two main ways in the mid-nineteenth century: as evidence of a "monogenic," or single, common origin for humans, or as evidence of a "polygenic," or disparate origin. As anthropologist Jon Marks writes, "Although Darwin settled the origins question in favor of monogenism, there was still considerable latitude in reconstructing the histories of the races. These histories could never be value-neutral, and here the entwined intellectual strands are indeed evident as alternative scientific explanations to a single biopolitical question: Why are there savages?"[3]

The intellectual history surrounding Chapter 5 of Darwin's *Descent of Man*, "On the Development of the Intellectual and Moral Faculties during Primeval and Civilised Times," is therefore quite complicated, and the ins and outs of early human evolutionary thought in the second half of the nineteenth century are too intricate for this introduction. Rather, in this chapter, I address Darwin's understanding of the diversity of humans through time, particularly with respect to their intelligence, their morality, and their invention of civilization. None of these three terms is appropriately defined by Darwin, which makes my discussion of his chapter a bit like an archaeological investigation, attempting to root out what formed his opinions, both scientifically and culturally. European thinkers of the nineteenth century used and misused Darwin's new ideas to discuss the "savages" they had met and "conquered," whereas American scientists began to generate a race-based biological anthropology, in part to morally justify the enslavement of Africans. A critique of Darwin's take on the races of humans follows in Chapter 7, written by Agustín Fuentes, but Darwin presents his ideas about humans' cultural evolution first.

In this essay, I excavate Darwin's main argument about cultural evolution through a close reading of the structure and language of Chapter 5 of *Descent* and then place the scaffolding of anthropology on top, in order to generate a better understanding of what Darwin knew and what we now know about intelligence and civilization.

What Darwin Said

The Influence of Alfred Russel Wallace

The epigraph that begins my essay is the first sentence of Chapter 5 of *Descent of Man*, which Darwin follows immediately with a reference to Alfred Russel Wallace's 1864 paper on the origin of human races, delivered at the Anthropological Society of London's meeting on March 1 of that year. This one-two punch is telling. Darwin cops to the limits of his explanatory abilities and vaunts Wallace's well-received paper, which was later published in *Anthropological Review*, an early anthropology journal. Before taking a close look at Darwin's words, then, it is worth investigating what Wallace (Figure 5.1)—who of course had also come up with the idea of evolution through natural selection—had to say seven years before *Descent*.

Much of Wallace's 1864 essay in *Anthropological Review* focuses on the then-current question of a monogenic versus a polygenic origin of humans, as well as the evolution of human culture, addressed in as

FIGURE 5.1. Alfred Russel Wallace, c. 1895. (Photograph by London Stereoscopic and Photographic Company. First printed in *Borderland*, April 1896.)

anthropological a way as possible for the time.[4] Wallace begins by largely espousing a monogenic perspective on human origins, one that reads much like contemporary anthropological understanding and fits with Darwin's perspective as well. The mid-nineteenth century was, of course, a period in which evolution through natural selection was becoming well accepted but the mechanism of heredity was still unknown. It was also a period in which human culture was assumed to be under the influence of natural selection in the same way that human biology was,[5] and this assumption comes through clearly in Wallace's essay: "Tribes in which such [high] mental and moral qualities were predominant would therefore have an advantage in the struggle for existence over other tribes in which they were less developed, would live and maintain their numbers, while the others would decrease and finally succumb."[6] Wallace has difficulty, however, reconciling humans' sociality and empathy with environmental checks on population, which would be necessary for his hypothesis about natural selection of mental and moral qualities to be borne out. Specifically, Wallace notes that even "in the rudest tribes, the sick are assisted at least with food" and that the weaker individuals "do not suffer the extreme penalty [death] which falls upon animals so defective."[7]

In order to further his argument in favor of the natural selection of human culture, Wallace tries his best to relate known primate qualities of sociality and empathy to humans' "intellectual and moral faculties." If social qualities evolve with mental capacity, he writes, "the better and higher specimens of our race would therefore increase and spread, the lower and more brutal would give way and successively die out, and that rapid advancement of mental organisations would occur, which has raised the very lowest races of man so far above the brutes, and, in conjunction with scarcely perceptible modifications of form, has developed the wonderful intellect of the Germanic races."[8]

Wallace, having taken as a given that physical adaptation in modern humans is less important than mental development, then doubles down on his idea of natural selection of both human intelligence and human culture and pushes it in an even more racist direction with what he considers proof of his thinking, as he rhetorically asks, "And is it not the fact that in all ages, and in every quarter of the globe, the inhabitants of temperate have been superior to those of tropical countries?"[9] Civilizations, Wallace avers, do not die out from one specific cause but rather are subject to "the inevitable effects of an unequal mental and physical struggle"[10]—that is, to cultural evolution. And who has reached the pinnacle of cultural evolution, according to Wallace? "The intellectual and moral, as well as the physical qualities of the European are superior," he concludes,[11] conflating cultural and physical differences and reifying white Western supremacy for his mid-nineteenth-century audience of educated white men. Wallace therefore believed that natural selection could operate on culture, albeit in a different way than it operated on biology, and wrote about the "inevitable extinction of all those low and mentally undeveloped populations with which Europeans come in contact."[12]

The audience's responses, however, to Wallace's monogenic take on how "races of man" were the result of natural selection of biology and culture made for a "chilly reception," according to historian Jeremy Vetter, who has traced Wallace's conflicts with both ethnologists and anthropologists. The 1864 paper was a turning point for Wallace, Vetter argues, as it presented new research material "under the strong influence of Herbert Spencer's social evolutionism as presented in *Social Statics*, which [Wallace] had recently read and also enthusiastically recommended to Darwin."[13] Wallace provided a copy of his paper to Darwin, insisting that he was on to something, even if the Anthropological Society of London disagreed.

As Darwin cites Wallace prominently in discussing "intellectual and moral faculties," his Chapter 5 can be seen as his attempt to grapple with Wallace's ideas nearly a decade later, while also admitting that, in the end, he cannot do the subject justice. Darwin's argument for cultural evolution and the evolution of civilizations therefore begins in Chapters 2 and 3 of *Descent*, with his ideas on morality and sociality, which are discussed by Brian Hare in his Chapter 3 commentary, earlier in this volume. Darwin ends Chapter 4 of *Descent* by asserting that "the early progenitors of man were, no doubt, inferior in intellect, and probably in social disposition, to the lowest existing savages; but it is quite conceivable that they might have existed, or even flourished, if, whilst they gradually lost their brute-like powers, such as climbing trees, &c., they at the same time advanced in intellect."[14] As an increase in intelligence is the key to raising oneself out of "brutishness," it is the topic to which Darwin turns first in Chapter 5 of *Descent*.

Natural Selection of Intelligence

From the outset, Darwin name-checks Wallace in his argument about the evolution of the human intellect, noting his agreement with Wallace that the "intellectual and moral faculties" of human beings "are variable; and we have every reason to believe that the variations tend to be inherited."[15] Humans, Darwin writes, have the ability to adapt their habits to new conditions of life: "He* invents weapons, tools, and various stratagems, by which he procures food and defends himself. When he migrates into a colder climate, he uses clothes, builds sheds, and makes fires; and, by the aid of fire, cooks food otherwise indigestible. He aids his fellow-men in many ways."[16] In these passages, we can see that Darwin was thinking about a biological foundation for human culture but that he also saw culture as constrained and brought about by both a natural and a built environment.

In order to justify his conclusion that intellectual powers are advanced through natural selection, Darwin writes:

> Relics of extinct or forgotten tribes have been discovered throughout the civilized regions of the earth, on the wild plains of America, and on the isolated islands in the Pacific Ocean. At the present day, civilized nations are everywhere supplanting barbarous nations, excepting where the climate opposes a deadly barrier; and they succeed mainly, though not exclusively, through their arts, which are the

* Darwin employs gendered language, as did almost everyone at that time, in speaking about humans and humankind. However, the use of male pronouns also emphasizes the patriarchal idea, current in the nineteenth century, that men were the prime movers of "civilization" and even that male primates and hominins were more important in the evolution of humans than were females. I preserve Darwin's and Wallace's language in quotations but update to gender-inclusive language elsewhere in this essay.

products of the intellect. It is, therefore, highly probable that with mankind, the intellectual faculties have been gradually perfected through natural selection; and this conclusion is sufficient for our purpose.[17]

Just as adult height is the product of both genetics and environment, Darwin says, culture—which he also refers to as "the arts" and links to the concept of civilization—is likewise the result of environment and inheritance. After he briefly covers the evolution of intelligence, Darwin quickly moves on to problematize it through the concepts of altruism, sympathy, and demographic change. I return to critique his idea of intelligence in a subsequent section.

Social and Moral Faculties: The Problem of Altruism

Instinctive feelings toward others, Darwin argues, are social qualities shared by a number of different animal species, acquired through natural selection and aided by "inherited habit."[18] Therefore, if two tribes of "primeval men" came into conflict, "a greater number of courageous, sympathetic, and faithful members, who were always ready to warn each other of danger, to aid and defend each other . . . would without doubt succeed best and conquer the other. . . . Thus, the social and moral qualities would tend slowly to advance and be diffused throughout the world."[19]

The immediate problem with this hypothesis, Darwin admits, is that "it seems scarcely possible . . . that the number of men gifted with such virtues, or that the standard of their excellence, could be increased through natural selection, that is, by the survival of the fittest."[20] Altruism, or the sacrifice of personal resources, even one's own life, without expectation of return, continues to be a difficult phenomenon to explain in evolutionary biology. Darwin believes that it exists because humans are constantly in search of praise—and loathe blame—by other people and that this is influenced by social instincts such as sympathy (covered in Chapter 3 of Descent). Concluding that early humans were influenced by praise and blame, Darwin suggests that "to do good unto others—to do unto others as ye would they should do unto you—is the foundation-stone of morality."[21] Humans' moral sense or conscience, he writes, is ultimately a complex sentiment, "having its first origin in the social instincts, largely guided by the approbation of our fellow-men, ruled by reason, self-interest, and in later times by deep religious feelings, [and] confirmed by instruction and habit."[22] Patriotism, fidelity, obedience, courage, and sympathy are therefore all the result of natural selection, in Darwin's mind.

Natural Selection as Affecting Civilized Nations: Demography and Checks on a Downward Tendency

Darwin also problematizes the idea of continuous improvement in human culture in his Chapter 5. He writes that "it is very difficult to form any judgment why one particular tribe and not another has been successful and has risen in the scale of civilization. . . . We are apt to look at progress as the normal rule in human society; but history refutes this."[23] Rather, "progress" is dependent on a number of factors, chief among them the environment. Darwin views the Inuit as highly ingenious, for example, for having developed numerous inventions to live in an Arctic climate, but also suggests that the severe climate has limited their progress. He disapproves of nomadism, calling it highly detrimental, and vaunts sedentism and property possession as well as agriculture as "indispensible requisites for civilization."[24]

But just as cultural advancement leads to civilization, in Darwin's mind, so too does civilization present an apparent demographic problem that doesn't square with his biological notion that the fittest will reproduce and succeed at higher rates. In drawing from the work of both Wallace and Francis Galton, Darwin's cousin and the man who coined the term "eugenics," Darwin compares and contrasts "civilized" and "savage" cultures:

With savages, the weak in body or mind are soon eliminated; and those that survive commonly exhibit a vigorous state of health. We civilised men, on the other hand, do our utmost to check the process of elimination; we build asylums for the imbecile, the maimed, and the sick; we institute poor-laws; and our medical men exert their utmost skill to save the life of every one to the last moment. There is reason to believe that vaccination has preserved thousands, who from a weak constitution would formerly have succumbed to small-pox. Thus the weak members of civilised societies propagate their kind. No one who has attended to the breeding of domestic animals will doubt that this must be highly injurious to the race of man. . . . Excepting in the case of man himself, hardly anyone is so ignorant as to allow his worst animals to breed.[25]

Here Darwin attributes humans' penchant for taking care of one another to his concept of sympathy and seems to speak against eugenics, writing that it is impossible for humans to go against their sympathetic instinct, "without deterioration in the noblest part of our nature. . . . Hence we must bear without complaining the undoubtedly bad effects of the weak surviving and propagating their kind."[26] After all, Darwin notes, one check against this may be that the weaker members of society do not marry as freely. On the other hand, Darwin married his first cousin, Emma Wedgwood, and was frequently concerned about the health of their ten children, fearing that their closely related union was "injurious."[27] Although only seven survived to adulthood, none of the children had evidence of problems due to inbreeding.

Even though Darwin presents the argument that rich and intelligent men succeed more often, and their children succeed more often, he also returns to Galton's findings regarding reproductive rates: "The very poor and reckless, who are often degraded by vice, almost invariably marry early, whilst the careful and frugal, who are generally otherwise virtuous, marry late in life. . . . Those who marry early produce within a given period not only a greater number of generations, . . . they produce many more children. . . . Thus, the reckless, degraded, and often vicious members of society tend to increase at a quicker rate than the provident and generally virtuous members."[28]

However, some checks exist to prevent a society from being overrun with Darwin's unsavory characters: "Malefactors are executed, or imprisoned for long periods, so that they cannot freely transmit their bad qualities. Melancholic and insane persons are confined or commit suicide. Violent and quarrelsome men often come to a bloody end. Restless men who will not follow any steady occupation—and this relic of barbarism is a great check to civilisation—emigrate to newly-settled countries, where they prove useful pioneers."[29] Natural checks to this "downward tendency," as Darwin calls it, also include a high rate of mortality among the "intemperate" men, as well as a high rate of mortality among women who marry at younger than twenty years of age and among both men and women who are "profligate."[30] Marriage in general is viewed as a civilizing force, and one that keeps both men and women—but mostly men—healthier and surviving to an older age. If the various checks Darwin specified "do not prevent the reckless, the vicious, and otherwise inferior members of society from increasing at a quicker rate than the better class of men," he concludes, "the nation will retrograde, as has occurred too often in the history of the world. We must remember that progress is no invariable rule."[31]

The Complexity of Human Culture

In spite of Darwin's problematic assumptions about capitalism, religion, and marriage being natural and the result of cultural progress, in rereading Descent to prepare to write this essay, I was surprised by how readily Darwin admits that the development of human culture is complicated and that its full understanding is beyond his ken, particularly the rise and fall of societies through time. "It is most difficult to say why one civilised nation rises, becomes more powerful, and spreads more widely than another, or why the same nation progresses more at one time than at another," Darwin writes. "We can only say that it depends on an increase in the actual number of the population, on the number of the men endowed with high intellectual and moral faculties, as well as on their standard of excellence."[32]

To illustrate his point, Darwin provides several historical and anthropological examples of cultural variation. The ancient Greeks, who were held in very high esteem by thinkers of his day, are a good example of a civilization with a higher intellect than any other culture that has ever existed, he asserts, but one that faltered. Because "natural selection acts only in a tentative manner, . . . development of all kinds depends on many concurrent favorable circumstances."[33] Whereas the Greeks "retrograded from a want of coherence between the many small states, from the small size of their whole country, from the practice of slavery, or from extreme sensuality," Western Europeans learned from this history and "now so immeasurably surpass their former savage progenitors and stand at the summit of civilization."[34]

Spain is another example Darwin provides. A culture that was once historically dominant suffered during the Inquisition, when "some of the best men—those who doubted and questioned, and without doubting there can be no progress—were eliminated during three centuries at the rate of a thousand a year."[35] On the other hand, Darwin attributes England's success as a colonial power to "daring and persistent energy," while also suggesting that "the wonderful progress of the United States, as well as the character of the people, are the results of natural selection; the more energetic, restless, and courageous men from all parts of Europe having emigrated during the last ten or twelve generations to that great country, and having there succeeded best."[36]

More precisely, Darwin focuses on an example of change in material culture over time to demonstrate his point about worldwide, human progress and the fundamental importance of independent discoveries:

> In all parts of Europe, as far east as Greece, in Palestine, India, Japan, New Zealand, and Africa, including Egypt, flint tools have been discovered in abundance; and of their use the existing inhabitants retain no tradition. There is also indirect evidence of their former use by the Chinese and ancient Jews. Hence there can hardly be a doubt that the inhabitants of these many countries, which include nearly the whole civilised world, were once in a barbarous condition.[37]

Darwin concludes his chapter by suggesting that the evolution of human culture is both part and parcel of human progress and, in his view, a transition from savagery to barbarism to civilization, an idea circulating among learned people of the time that would be most famously written down a few years later by American anthropologist Lewis Henry Morgan in his 1877 book *Ancient Society*.[38] "Man has risen," Darwin ends, "though by slow and interrupted steps, from a lowly condition to the highest standard as yet attained by him in knowledge, morals, and religion."[39]

Darwin wrote *Descent* at what proved to be an extremely important time for modern science. As anthropologist Jon Marks notes in the conclusion to his essay on the history of human origins and variation in the nineteenth century, "Darwinism emerged entangled with revolutionary scholarly approaches to human diversity and to the Bible itself, as 19th century biology, anthropology, and theology all wrestled with their newly emerging paradigm."[40]

What Do We Know Now?

In retrospect, most of Chapter 5 of Darwin's *Descent* will give any social scientist pause. Given his employment of patriarchal language, his conflation of religion with morality, and his uncritical naturalizing of the Western European and colonialist way of life, it is relatively easy to poke holes in Darwin's explanation of cultural evolution and civilization. While Brian Hare critiques Darwin's concepts of morality, sociality, and sympathy in Chapter 3 of *Descent*, it is worth exploring here what we now know about human intelligence and about the rise and fall of civilizations.

Brain Size, Complexity, and Intelligence

Darwin notes in Chapter 4 of *Descent*, on the evolution of the human body from a primate ancestor, that the human brain is large relative to body size and that the relationship between brain size and "mental faculties" is clear. Unfortunately, Darwin also cites the scientific understanding of his day, which held that, "there exists in man some close relation between the size of the brain and the development of the intellectual faculties, . . . supported by the comparison of the skulls of savage and civilised races, of ancient and modern people, and by the analogy of the whole vertebrate series."[41] He cites as his sources of information people like Paul Broca and Johann Friedrich Blumenbach, the latter of whom is perhaps best known for creating and popularizing the five-fold grouping of human races, and who held that Adam and Eve were Caucasians and every other race was produced by degeneration. Both Broca and Blumenbach were pioneers in cranial anthropometry (skull measuring), which ultimately formed the basis for scientific racism in early biological anthropology, when physical traits were used to justify cultural and structural violence, including slavery, patriarchy, and colonialism.

Three major issues with Darwin's use of intelligence in explaining culture and civilization are: 1) his assumption that brain size is a proxy for intelligence, 2) his assumption that intelligence can be quantified, and 3) his inference that quantifications of intelligence reflect heritable and immutable traits in humans. These assumptions can in part be traced to the works of his aforementioned cousin, Francis Galton, a prolific polymath. In the field of statistics, Galton figured out how to quantify the normal distribution, or bell curve, with standard deviation; he invented the regression line r for representing the correlation coefficient; and he coined the term "regression to the mean." In forensics, Galton was the first to study heritability in fingerprints and to work out the statistical likelihood that two people would have the same pattern.

Francis Galton was also intrigued by his cousin Charles Darwin's ideas on variation within species and sought to measure human variation in both biology and behavior, or, as he called it, "nature and nurture." This latter desire led Galton to measure "desirable" human qualities and to attempt to figure out which were heritable traits, a research plan that Darwin himself disagreed with. Galton devised one of the first known intelligence tests, suggesting that human intelligence could be objectively quantified through measuring such things as head size, vision and hearing accuracy, and reaction time. Following Darwin's death in 1882, Galton began calling his research into those people who were "born well" by the Greek-based neologism "eugenics."

Our contemporary understanding of intelligence is based on equally flawed but more recent tests devised in the first half of the twentieth century. Measuring what is widely known as IQ, or intelligence quotient, a term that dates back to 1912, the earliest modern test was developed in France by Alfred Binet and later became the Stanford-Binet scale still in use today. Binet, attempting to identify children who might need scholastic intervention in order not to fall behind their peers in their age cohort, developed a test with questions about recognition of standard objects, verbal definitions, execution of simple commands, and working memory tests. A child's IQ under this scheme was his or her ratio of tested mental age to expected chronological age.

Around the same time, psychologist Henry Goddard began to use the IQ test to classify people with intellectual challenges; his terms "moron" (51–70 IQ), "imbecile" (26–50 IQ), and "idiot" (0–25 IQ) were used by psychologists and governments for decades. A true eugenicist, Goddard believed that low-IQ individuals should be removed from society by institutionalization or sterilization. Goddard backed up his eugenics with a healthy dose of racism and classism; in a study he ran on immigrants to New York's Ellis Island, he found that roughly 80 percent of those in steerage class were "feeble-minded."[42]

The most famous scientific criticism of IQ testing and the general quantification of human intelligence came in 1981, with the publication of Stephen Jay Gould's *The Mismeasure of Man*, in which Gould tackled the phenomenon of intelligence quantification and traced its ill effects on human society. Gould called the problematic concept "biological determinism" and defined it as the idea that "shared behavioral norms, and the social and economic differences between human groups—primarily races, classes, and sexes—arise from inherited, inborn distinctions and that society, in this sense, is an accurate reflection of biology."[43] He sought to counter biological determinism by laying bare the ideological biases behind craniometry and psychological testing.

Gould specifically saw—in his 1981 edition of *Mismeasure of Man* as well as in his 1996 update after the publication of Richard Herrnstein and Charles Murray's *The Bell Curve: Intelligence and Class Structure in American Life*—an unfortunate resurgence in the popularity of biological determinism, which once undergirded the practice of eugenics. Gould wrote that biological determinism is correlated with "episodes of political retrenchment, particularly with campaigns for reduced government spending on social programs, or at times of fear among ruling elites, when disadvantaged groups sow serious social unrest or even threaten to usurp power."[44] While Darwin suggested that there were natural limits on the "downward tendency" of the human population, Gould a century later identified more specific political mechanisms that have lasted well into the twenty-first century. Thus, in *The Mismeasure of Man*, Gould criticized men such as Paul Broca who were engaged in scientific racism and whose work was based on culturally biased, *a priori* expectations.

At the heart of what Gould critiqued was not only biological determinism but also, particularly in his response to *The Bell Curve*, the unstated assumptions held by many white Westerners that make IQ testing so insidious. Gould isn't alone. Data scientist Eric Siegel has summarized Herrnstein and Murray's book and its implicit racism by noting, "*The Bell Curve* endorses prejudice by virtue of what it does not say. Nowhere does the book address *why* it investigates racial differences in IQ. . . . The net effect is to tacitly condone the prejudgment of individuals based on race."[45] Psychologist Howard Gardner, famed for his work on multiple intelligences, further avers that the rhetoric in *The Bell Curve* encourages readers to align themselves with extreme positions— such as the abolition of affirmative action and the curbing of reproduction by people with low IQs.[46]

Darwin clearly shares with Herrnstein and Murray these *a priori* assumptions about intelligence and similarly conflates intelligence with socioeconomic superiority, health, and well-being in the form of longevity, civilization, and progress.[47] They're not the only men to assume racist perspectives on intelligence. In 2007, James Watson, the famous American molecular biologist who won the Nobel Prize for his work with DNA, told a reporter that "all our social policies are based on the fact that their [Africans'] intelligence is the same as ours—whereas all the testing says not really."[48] While Herrnstein and Murray, Galton, and others may have found data to support different IQs in different racial groups, those data are meaningless and cannot be interpreted without social and economic context. It's the context—the socioeconomic hardships, the discriminatory health-care system, the lack of access to education, the criminal justice system—that social scientists have shown to be impossible to separate from any measure of human intelligence, as inequality is a cultural construct, not a natural, biologically deterministic pattern.

The idea of major differences in intelligence by race and/or social class, although thoroughly debunked, has reemerged in the United States, England, and elsewhere in the past few years. Although Gould passed away in 2002, his warnings about "political retrenchment and destruction of social generosity"[49] are unfortunately once again relevant amid the resurgent popularity of biological determinism and white nationalism.

Civilization and Progress

Just as scientific racism surrounding intelligence belies its proponents' assumptions and prejudices about other groups, the contemporary discussion of civilization and progress has similar themes. Take former U.S. congressman Steve King, a man whose fifteen-year history of racist polemics, specifically surrounding civilization and birth rates, has warranted its own time line at the *New York Times*. In 2002, King tried to get a bill passed requiring public schools to teach that the United States is the greatest nation in the world, thanks to Christianity and Western civilization. In 2011, King spoke out against hormonal birth control, noting, "If we let our birthrate get down below the replacement rate, we're a dying civilization." And if there's any question that King was speaking specifically about the white birth rate, in 2017 he wrote on Twitter that "we can't restore our civilization with somebody else's babies." In early 2019, King told a reporter, "white nationalist, white supremacist, Western civilization—how did that language become offensive?"[50]

King is right that "civilization" is not a universally offensive term; but it has always been a loaded and exclusionary one. That language came about during the Enlightenment, in the eighteenth century, when "civilization" as a term was borrowed from the French as a sort of politically correct replacement for the English word "civility," the opposite of "savagery."[51] By the nineteenth century, "civilization" quickly came to mean the idea of human cultural development or linear progress and was therefore something to attain but also something to force on those peoples who were not considered to be civilized, to teach them to participate in the Western, colonialist economy. We get definitions like this one from American essayist Ralph Waldo Emerson in 1862: "A certain degree of progress from the rudest state in which man is found,—a dweller in caves, or on trees, like an ape, a cannibal, an eater of pounded snails, worms, and offal,—a certain degree of progress from this extreme is called Civilization."[52]

By the early twentieth century, social thinkers had largely jettisoned the idea of progress but still sought to identify reasons for culture change. For example, in 1913, the notable sociologists Emile Durkheim and Marcel Mauss wrote "Note on the Notion of Civilization." Admitting that nations are complex and difficult to study, as shown by archaeologists who attempted to make sense of past material culture, Durkheim and Mauss write that civilizations are "systems of facts that have their own unity and form of existence." In their view, we could define a Christian civilization, a Mediterranean one, or a Northwest American one. "A civilization constitutes a kind of moral milieu encompassing a certain number of nations, each national culture being only a particular form of the whole," they write. Durkheim and Mauss therefore find the identification of successive stages in cultural progress to be inadequate to understanding the social organism that is civilization. "There have been and there still are diverse civilizations," they note, and "one could ask what are the diverse conditions which determine variations in the areas of civilizations; why have they stopped here or there; what forms have they taken and what factors determine these forms."[53] The notion of civilization was therefore decoupled from the idea of progress by early twentieth-century sociologists, and both terms became ideas worthy of academic research.

It wasn't only sociologists examining culture either. Evolutionary biologist Julian Huxley, writing in 1955, notes that extending the idea of evolution to culture is problematic because of "the erroneous idea that biological evolution could be represented by a single straight line of inevitable progress, and the . . . conversion of this into an evolutionary strait-jacket for culture."[54] By the mid-twentieth century, the idea of cultural evolution had become popular again, and the natural environment was cited as a limiting factor in the formation of human societies. Huxley, for example, writes that "the environment does not determine the culture, but does condition and may limit it—for instance, through the extremes of climate, or the prevalence of debilitating disease."[55] This is not far removed from Darwin's ideas in *Descent*, in which he writes, "The Esquimaux, pressed by hard necessity, have succeeded in many ingenious inventions, but their climate has been too severe for continued progress."[56]

This environmentally deterministic perspective on the rise and fall of civilizations has been eagerly embraced well into the twenty-first century by such science writers as Jared Diamond and Yuval Noah Harari.[57] However, the researchers whose study focuses on cultural change over time—archaeologists—have a different understanding, and their work on the rise and fall of complex societies complicates the picture considerably.[58] Research over the last several decades has shown that identifying one "prime mover" is impossible; rather, past activity is diverse, and circumstances are unique to a specific culture or society. Archaeologists caution that Diamond, Harari, and others over-essentialize the processes that lead to the rise and fall of past and present societies in order to write a tidy—but inaccurate—story of the past.

The way that archaeologists today discuss the rise and fall of human societies is first by defining terms. We no longer subscribe to the nineteenth-century ideas of Lewis Henry Morgan and others who insisted that humans progress linearly from savagery to barbarism to civilization. Rather, we talk about different forms of societies in a more diversified way, influenced by Franz Boas and other early twentieth-century anthropologists who drew from sociological thought. "Complex societies" are formed by several different aspects of culture—cities, agriculture, inequality, surplus, specialization, politics, and economics—that are integrated and work together.[59] This definition of cultural complexity was formed by archaeologists in the late twentieth century but has stuck around as a shorthand for the phenomenon we see in the present day and in many societies of the past. However, most archaeologists today would agree with Robert Chapman, when he notes in his book *Archaeologies of Complexity*:

> If we turn to the long-term record of archaeology, there is, I think, no doubt that the human societies which inhabit this planet have become more complex (in the sense of interconnectedness) and more unequal, both within individual societies and at the level of global relations. This is a gross trend, superimposed on shorter-term records of evolution and devolution, of "rise" and "fall" of more complex societies such as the earliest states, of change at different rates and scales, or to put it more grandly, of history. There have been many different forms of society, as there are today, and complexity should not be conceived as the ultimate goal of social evolution.[60]

The archaeology of complexity that contemporary researchers engage with is not as simple as what Darwin wrote of in the nineteenth century nor the same as what Mauss and Durkheim noted in the early twentieth century. Rather, culture change is addressed according to one or more archaeological theories, each offering a perspective that helps archaeologists identify and explain past behavior of individuals and cultures. These theories—neoevolutionism, historical particularism, and practice theory—help archaeologists address artifacts used in a complex society, the power and ideology behind the rise of the society, and how individual and collective action worked together to form it. Use of these theories requires archaeologists to be reflexive and to try to eliminate their own cultural biases when interpreting the remains of the past. But since we are all a product of our cultural upbringing, sometimes archaeologists slip up and reveal their biases. For example, Chapman notes that archaeologist Ian Hodder, regarding his well-known research at Çatalhöyük in Turkey, writes about a "low degree of social complexity," and this, for Chapman, "raises the question as to how far such notions of social evolution are embedded in everyday thought and action in Western society"[61]—today, as they were in Darwin's time as well.

If archaeologists' assumptions about the "rise of civilization" have yet to be fully examined, it is fair to say that those ideas regarding "collapse" have not either. Collapse or fall of a complex society is well known to archaeologists who have studied the Roman, Ottoman, Inca, and Mali Empires, the Maya civilization, the Han dynasty, and many others. Collapse may come when there are drastic changes to parts of the cultural system that cannot be easily overcome. For example, encroaching war can obviously cause collapse, but population pressure, a pandemic disease, or climate change that affects the society's food supply can as well. Perhaps a society faced with these pressures is able to create new institutions to deal with them—if it is unable to, however, the entire system can be brought down.

Given the Western cultural assumption that there is constant forward progress, and given the lack of collective memory of societal transformation, many people today tend to believe that their society is the norm and that it is not going anywhere. This widely held view also correlates with "civilization" and with cultural "progress"—and today these ideas have been used quite often by North American and European politicians to pit "us" (white upholders of the Western, Judeo-Christian tradition) against "them" (nonwhites, immigrants, non-Christians, and others). U.S. congressman Steve King may be a notorious and unrepentant racist, but his spoken and unspoken views on "civilization" and cultural "progress" are widely held among twenty-first century white Americans, just as these same ideas were widely held by Europeans in Darwin's time.

Where Do We Go Now?

The increase, decrease, stasis, and collapse of human cultural complexity is an ongoing research topic of interest within the field of anthropology. Whereas human biological evolution is far better understood today than it was in Darwin's time, thanks primarily to the discovery of genes and the burgeoning of that field of research, human societies are still messy. We can explain biological characteristics of humans through natural selection, mutation, genetic drift, and gene flow, but attempts to do this with culture have not borne the same fruit, as there is very little variation within cultures that can be seen as "competition" in the

Darwinian sense, and therefore there is no analogy for the "struggle for existence." And while a Darwinian approach has been attempted in measuring and explaining human intelligence, anthropology and other social scientific fields have similarly shown that the results of that approach are related in no small part to cultural correlates such as inequality rather than assumed biological correlates such as brain size.

What anthropology has taught us, since the time of Darwin and Wallace, is that the Enlightenment ideal of human progress and the capitalism and inequality arising from the industrial revolution that underlie many early theories about human society are not "natural" nor are they necessarily something all humans on the planet should strive for. As the eminent author and lapsed anthropologist Kurt Vonnegut once commented:

> I didn't learn until I was in college about all the other cultures, and I should have learned that in the first grade. A first grader should understand that his or her culture isn't a rational invention; that there are thousands of other cultures and they all work pretty well; that all cultures function on faith rather than truth; that there are lots of alternatives to our own society. Cultural relativity is defensible and attractive. It's also a source of hope. It means we don't have to continue this way if we don't like it.[62]

As more academics and scientists wade into social media, attempting to convince a growing number of doubters about the importance of vaccinations, climate catastrophe, historical facts, and biological discoveries, it becomes clear that many of us are trying to stem a rising tide of white nationalism barely disguised as "cultural heritage." If politicians and voters, however, fail to pay attention, the complex, global society we live in—our current civilization—may falter.

6

Ranking Humanity among the Primates

John Hawks

If man had not been his own classifier, he would never have thought of founding a separate order for his own reception.

CHARLES DARWIN, *DESCENT OF MAN*, CHAPTER 6[1]

ONE OF CHARLES DARWIN'S great insights was that the relationships of living organisms form a tree. But where do humans fit in the great tree of life? In Chapter 6 of *The Descent of Man*, "On the Affinities and Genealogy of Man," Darwin presented his ideas about the relationships of humans to other living groups of animals. Darwin accepted that humans fit with monkeys and apes into the group known as primates. He reviewed the comparative anatomy of this group, drawing upon more extensive treatments by other naturalists, including Charles Lyell, Thomas Henry Huxley, and St. George Mivart.[2] He emphasized those aspects of anatomy and behavior that supported his reasoned arguments for the common descent of living creatures.

On three topics, Darwin staked claims that diverged from the beliefs of many of his contemporaries. One was the position of humans in biological classification. Most naturalists of the time, even those who recognized common descent, nonetheless classified humans as a group apart from nonhuman apes and monkeys. Darwin argued that humans fall within the "anthropomorphous" apes and that a natural classification must be based upon genealogical relationships, not novel features found in humans alone. A second topic was Darwin's insistence upon the value of nonadaptive traits for identifying relationships. Darwin acknowledged that adaptive features like upright walking and large brains were important to humans, but he argued that vestigial or apparently useless traits are more reliable for understanding our affinities. Third, Darwin took a step beyond any previous scientist with his novel biogeographic hypothesis that humans had originated in Africa. It is for the last of these that Darwin won his current reputation as an oracle of human origins. Against the prejudices of his time, he recognized that humanity began not in Eden, nor in Asia, but in Africa.

In the last few pages of Chapter 6, Darwin turns to a more speculative area: the "lower stages"[3] in the genealogy of humans. It is here that Darwin came closest to a fish-to-human account of our evolutionary history. Looking into the past beyond the dawn of humans, his text turned attention for the first time to even earlier progenitors: once as apes, before that as marsupials, ultimately as lancelets.

For these reasons, this chapter has great resonance in the history of anthropology, cited more in the study of human evolution than any other part of Darwin's work. Naturally, 150 years of advances in scientific methods and evidence have changed our picture of where humans fit among the primates. Biologists have explored the diversity of the primate order within habitats around the world, adding hundreds of species never described in Darwin's time. The fossil record of primates has burgeoned with discoveries representing many of the key lineages of living primates, as well as many extinct groups, across more than 60 million years of evolutionary time. During the past fifty years, genetics has provided a new line of evidence about primate relationships, resolving many long-standing questions and raising new ones. These discoveries have resolved our ancestral connection to other primates and mammals, even as our attempts to classify humans have remained contentious.

Darwin's Primate Evolutionary Tree

To understand how today's view of our place in nature compares to Darwin's, we must go beyond the text of *Descent* for a picture of Darwin's primate tree. Modern biologists examine the "affinities and genealogy" of our species by drawing trees. Yet, in all of his published work, Darwin included only one single image of a tree. No trees appear in *Descent* at all. Throughout his discussion of the place of humans among the vertebrates, Darwin did not even use the word "tree."

Darwin's notes and correspondence reveal trees for the living and fossil primates that are lacking in his book. These greatly clarify how Darwin constructed his views and where he drew ideas from other naturalists. The historian of biology J. David Archibald reviewed Darwin's primate tree sketches and their role in Darwin's thinking.[4] One of these sketches, with the date April 21, 1868, plots an arrangement for humans, orangutans, chimpanzees, and gorillas, as well as lemurs, New World monkeys, and Old World monkeys (Figure 6.1). This tree diagram from Darwin's notes corresponds closely to his description of primate relationships in *Descent*. Darwin placed the "anthropomorphous" apes—what we today call the great apes—upon a single branch with a three-way split between the chimpanzees and gorillas, the orangutans, and *Hylobates*, or gibbons. The stem of this branch also gives rise to a second branch leading to humans. At first Darwin labeled this "*Homo*," then scratched out the Latin term to replace it with "Man." Deeper in the tree, this branch connects to another leading to the Old World monkeys, with the labels "*Cercopithecus*," "*Macaca*," "Baboon," and "*Semnopithecus*." Even deeper, a branch leads to the New World monkeys, and at the deepest point in the tree, a branch leads to "Lemuridae." At the very base of the tree is the label "Primates."

Darwin drew his information on primate relationships from other contemporaries, many his close correspondents, and it is reasonable to see his idea as a summary of all the data available in his time. Some of his ideas came from Thomas Huxley's *Evidence as to Man's Place in Nature* and Charles Lyell's *Geological Evidences of the Antiquity of Man*, both published in 1863 and both treating primates at substantial

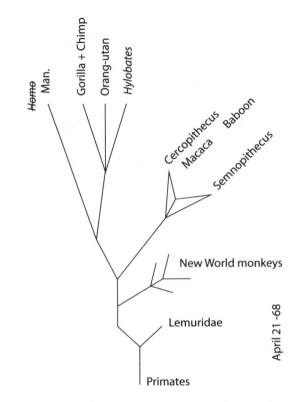

FIGURE 6.1. Tree of primate relationships, redrawn after Darwin's manuscript
sketch (MS DAR80.b91r, Cambridge University Library). Aside from grouping
the gibbons and great apes together apart from humans, this tree fits the
present understanding of relationships among the primates.

length. Further information came from St. George Mivart, who focused
on the anatomical relations of primates and included anatomical trees
in his work, as well as the German naturalist Ernst Haeckel. Haeckel in
particular became famous for his evolutionary trees, coining the term
"phylogeny" for the treelike arrangement of organisms.

Some of the most up-to-date discoveries for Darwin were from the
fossil record. At the time, the most well-known primate fossil was *Dryopithecus fontani*, from Saint-Gaudens, France, described by Édouard
Lartet in 1856.[5] Haeckel had considered this fossil ape as a possible in-
termediary between apes and humans, an idea that would be shared by
some later scientists but is not accepted today.[6] Darwin did not go so

far. He accepted Lartet's assessment of *Dryopithecus* as gibbon-like in many ways, not a close relative of humans. He found the central aspect of *Dryopithecus* to be its age: evidence of an anthropoid ape in "the Upper Miocene"[7] implied an even greater antiquity for human origins, perhaps "at an epoch as remote as the Eocene* period."[8] The other fossil primate Darwin considered was *Mesopithecus pentelicus*, discovered in Greece by Jean Albert Gaudry.[9] Gaudry considered it a possible ancestor linking the two branches of cercopithecoid monkeys, which we know today as cercopithecines (baboons, macaques, guenons, and vervet monkeys) and colobines (leaf-eating langurs and colobus monkeys). Darwin saw *Mesopithecus* as support for his precept that gaps between present-day animals were produced by extinction of intermediate forms.

Primate Relationships Today

More than 500 species of living primates, belonging to more than eighty genera, are known to science today. Both these numbers have recently been increasing. Geneticists and field biologists continue to refine their knowledge of the diversity of natural populations in many parts of the world, identifying species that remained unrecognized in the past. Natural geographic ranges of primates include parts of Africa, Asia, Europe, and North and South America, as well as island Southeast Asia, Japan, Taiwan, and Madagascar. Evidence of relationships from DNA has helped to establish the pattern of branching as well as the approximate times that different groups diverged from one another.[10] Fossil evidence has added to this picture, providing a hard record of the first appearances of many primate groups, as well as documenting their arrival in various parts of the world.

The last common ancestors of today's primates lived shortly after the beginning of the Paleogene period, some 66 million years ago. Living and fossil primates share many traits that are the legacy of these common ancestors. Binocular vision, grasping hands with opposable

---

* The Eocene epoch is now known to have lasted from 56 to 33.9 million years ago. The earliest possible hominin fossils are around 6–7 million years old.

thumbs, and broad fingertips with nails instead of claws are shared by primates as diverse as humans and galagos. These traits are adaptations that enabled the common ancestors of all primates to thrive in trees.

Like other mammals, primates emerged from the shadows after the extinction of dinosaurs, rapidly diversifying into a multiplicity of forms. Several families of these early primates evolved during the Paleocene and Eocene, before 35 million years ago, and became extinct in later time periods. Fossils from western North America, China and Southeast Asia, Europe, and North Africa document these early primates. Geneticists can examine their emergence only indirectly, through the DNA similarities of their living descendants. Still, genetic evidence has allowed for estimation of the times when living groups diverged and has overturned many ideas about relationships that were based on fossils and anatomy of living primates.

The two deepest branches of living primates are a good example of the interplay of these areas of evidence. Modern scientists recognize these deepest primate branches as the Strepsirrhini and Haplorhini, the names describing the forms of their noses. The moist-nosed strepsirrhines include lemurs, lorises, and galagos, while the haplorhines are the tarsiers, monkeys, and apes, including humans. During the past thirty years, this deepest split in the tree has become near universally accepted, but earlier scientists had different views about the deepest branches. Nearly all agreed that lemurs and lorises share a common heritage, reflected in traits such as a "toilet claw" and a "tooth comb" of closely packed lower incisor teeth, both used in grooming the fur. Scientists also agreed that monkeys, apes, and humans are a natural branch, historically known as "anthropoid primates" and today classified as Simiiformes. Paleontologists trace these two groups back to two successful and diverse groups during the Eocene, the lemur-like adapids (Adapidae) and the tarsier-like omomyids (Omomyidae).

This primary division, between strepsirrhine and haplorhine primates, hinges on the placement of tarsiers—small nocturnal primates from the Philippines, Sulawesi, and other islands of Southeast Asia. European naturalists knew of tarsiers in the eighteenth century but varied greatly in how they classified them—at first as monkeys, sometimes as non-primates.[11] Carolus Linnaeus classified tarsiers within his monkey genus *Simia*, while Étienne Geoffroy Saint-Hilaire recognized tarsiers as similar to lemurs and galagos.[12] Darwin's contemporaries Mivart and Haeckel placed tarsiers with lemurs, a view that would persist through most of the twentieth century.[13] Yet early in the twentieth century an alternative view emerged that tarsiers might indeed be closer to monkeys, apes, and humans—the Simiiformes—than to strepsirrhines.[14] This debate continued through the discovery of fossil omomyids, many of which seemed similar to tarsiers. Were the first ancestors of monkeys, apes, and humans to be found among the tarsier-like omomyids or among the lemur-like adapids? Late twentieth-century discoveries in Eocene fossil deposits in Asia of early tarsier-like simiiform primates such as *Eosimias* provided support for the hypothesis that tarsiers and simiiforms shared a common ancestry.[15] Genetic data from tarsiers, most recently whole-genome sequences, have confirmed that these primates are the closest living relatives of the monkeys and apes.[16]

The position of tarsiers serves as one of many examples demonstrating that the deep relationships of today's groups are not those envisioned by Darwin and his contemporaries. To his credit, Darwin recognized a "break" between tarsiers and other lemurs, comparing it to the break "between the orang and its nearest allies."[17] Indeed, today's primate phylogeny places orangutans in a very different position from the one it held in Darwin's tree. Whereas Darwin placed all the apes on a branch opposed to the human line, today's ape phylogeny has at its base a division between the gibbon and great ape branches, and then within the great apes between orangutans and the African ape-hominin branch. It is notable that even though Darwin tended to lump the "anthropomorphic apes" together—just as Huxley, Lyell, and Mivart had done—he nonetheless described the African apes as "man's nearest allies" (further discussed below).[18]

In the modern phylogeny (Figure 6.2), humans and the great apes are part of a biological family known as Hominidae, linked with gibbons and siamangs into the superfamily Hominoidea. We are catarrhines, just

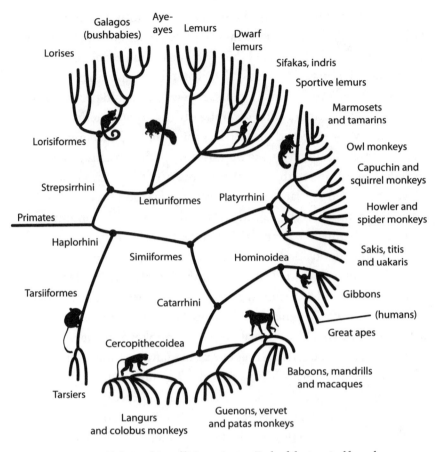

FIGURE 6.2. Relationships of living primates. Each of the terminal branches
of this tree corresponds to a genus of living primates, numbering eighty-one in all.
This tree relies upon the phylogeny presented by Christian Roos and Dietmar Zinner,
"Primate Phylogeny," *The International Encyclopedia of Primatology* (2016): 1–5, on
the basis of recent genetic and morphological comparisons.
(Primate silhouettes from PhyloPic [CC-BY])

as Darwin asserted, together with the monkeys of Africa, Asia, and Europe. The platyrrhine monkeys of North and South America belong to their own branch, separated from the catarrhines sometime prior to the Oligocene epoch, more than 35 million years ago. We now understand primate biogeography to be the result of ancient dispersals and founder events, including the early transatlantic dispersal of platyrrhine ancestors to South America, one and possibly two founder events bringing lemurs and aye-ayes to Madagascar, and the Miocene dispersal of apes across Eurasia. The biogeography produced by these ancient founder events was very important to Darwin's thinking and would lead him to speculate on the origins of the human lineage.

An African Origin

On the Birthplace and Antiquity of Man.—We are naturally led to enquire, where was the birthplace of man at that stage of descent when our progenitors diverged from the Catarhine [sic] stock?

CHARLES DARWIN, DESCENT OF MAN, CHAPTER 6[19]

The science of human origins has a peculiar obsession with cradles. This fascination did not begin with Darwin, of course. In the Judaic, Christian, and Islamic traditions, the idea of a "birthplace" for humanity goes back to scriptural stories, and most human cultures have their own origin stories. Darwin was not even the first to consider this question from the standpoint of transformation. In 1775, Johann F. Blumenbach formulated a theory that environments had transformed human groups after their creation, which led him to the conclusion that the least transformed group, "Caucasians," must have remained close to the Garden of Eden.[20]

Darwin's innovation was to look beyond humanity's root. Humans share common ancestors with other primates, and the place where our own early ancestors lived is likely to be the same as theirs. Humans are related to catarrhines and more specifically to the African apes. Taking this as a starting assumption, Darwin ruled out Australia, where there are no living nonhuman primates at all, and the Americas, where only the platyrrhine monkeys exist. He stated that gorillas and chimpanzees are "man's nearest allies," making it "somewhat more probable that our early progenitors lived on the African continent than elsewhere."[21]

Yet Darwin recognized that the mere fact that our relatives can be found in Africa today does not mean they always lived there. Here Darwin again turned to *Dryopithecus*, the extinct European ape that lived in France, where no apes are found today. "Since so remote a period," he wrote, "the earth has certainly undergone many great revolutions, and there has been ample time for migration on the largest scale." With such great changes, he concluded, "it is useless to speculate on the subject."[22]

For nearly a hundred years, anthropologists have developed the fossil record of human relatives in Africa (see Chapter 4 of this volume). From the first discovery of an early hominin in Africa up to the present day, paleoanthropologists have approvingly cited Darwin's ideas of an African origin for our species, from Raymond Dart and Louis Leakey to Tim White and colleagues.[23] The best current knowledge is that early hominins such as *Australopithecus* were upright, bipedal apes that lived only in Africa, while *Homo* dispersed from an African source into other parts of the world. Modern humans represent one of many burgeoning populations to have dispersed originally from an African source population. The late Phillip Tobias translated the science into a memorable sentiment: "Africa gave the world humanity, and that is no small thing."[24]

Still it would be premature to say that the deeper origins of the hominin lineage are settled. The record of ape evolution in the Miocene of Africa is rich, but paleontologists have not yet found fossils representing the last common ancestors of humans, chimpanzees, and gorillas. Some scientists suggest that the fossil apes of Europe, some of them contemporaries of *Dryopithecus*, might provide clues about the ancestry of the African apes and humans. David Begun and coworkers, for example, pointed to European apes such as *Dryopithecus*, *Ouranopithecus*, and *Rudapithecus* as the earliest-known members of the branch leading to both living African apes and humans.[25] They posed the possibility that one of these Eurasian apes may have migrated into Africa, giving rise to the gorilla, chimpanzee, and hominin lines. These researchers, too, approvingly quote Darwin—but instead of focusing on the African origin, they draw attention to Darwin's mention of *Dryopithecus* as evidence of the "ample time for migration on the largest scale."[26] Nevertheless, for now, as discussed in Chapter 4, the earliest possible hominins— *Sahelanthropus*, *Orrorin*, and *Ardipithecus*—have come from fossil localities in central and eastern Africa.

The Taxonomic Rank of Humanity

Modern biological systematists recognize groups of organisms on the basis of common descent. Indeed, "systematics" today is the synthesis of taxonomic classification and phylogenetics. This is one of Darwin's strongest legacies in biology. Using genealogy as a basis for classification is a theme he takes up in his Chapter 6, following from extensive discussions in Chapter 13 of *On the Origin of Species* (1859) and again in *Variation of Plants and Animals under Domestication* (1868). In *Descent*, his immediate concern is where humans should rank in natural classification.

The modern classification aligns with primate phylogeny. Humans are a species, *Homo sapiens*. Our closest living relatives are chimpanzees and bonobos, today classified within the genus *Pan*. Fossil discoveries of the past 150 years have unearthed many extinct relatives of humans that are closer to living people than to any other living primate. This branch of the primate phylogeny, leading not only to living people but also to extinct forms such as *Homo erectus* and *Australopithecus afarensis*, is recognized as the biological tribe Hominini, and its members are known as hominins. Gorillas, chimpanzees, and bonobos are linked with hominins as a biological subfamily known as Homininae. Adding the branch leading to orangutans and their extinct relatives, the great apes and humans together make up the family Hominidae and are known as hominids. The taxonomic rank of humans in this system is that of a biological tribe, below the subfamily and above the genus.

In discussing the proper rank of humans in biological classification, Darwin faced a proliferation of names for primate groups. Each name reflects a different set of assumptions about anatomical similarities. The names used in the modern classification of apes and humans can be traced to John Edward Gray, who in 1825 formulated a system to classify the biological specimens at the British Museum (later the Natural History Museum).[27] Systematists have shifted the meanings of the terms from Gray's original definitions. Gray's family Hominidae included not only humans and chimpanzees (gorillas were not yet known to science in 1825) but also other catarrhine primates. Gray's Hominina included only humans. Mivart used Hominidae in his classification, opposing it to Simiidae, which included all nonhuman catarrhines, and Darwin refers to Hominidae in a footnote.[28] Darwin's text discusses Huxley's 1869 system, with Anthropidae for humans and Simiadae for all monkeys and nonhuman apes.[29] Lyell retained the distinction between the two orders Bimana, humans, and Quadrumana, nonhuman primates.[30] This scheme had almost as long a pedigree as Linnaeus's concept of the order Primates, having been introduced by Blumenbach and elaborated by Georges Cuvier.[31] Haeckel employed the classification established by Geoffroy in 1812, separating Catarrhini and Platyrrhini, and including humans with apes as Anthropoides within Catarrhini.[32] This was truly a tangled record of classification, with each specialist drawing from different prior sources for different purposes. Darwin's discussion drew from all of them. He argued that humans rank within the catarrhines, not at the level of an order, as would be designated by Bimana. Yet, even so, the rest of the book reflects a basic confusion, as Darwin relied upon the term Quadrumana throughout.

Does rank matter? A ranked hierarchy of kingdom, class, order, genus, and species was part of Linnaeus's original scheme, retained today with even more levels. But the tree of life has thousands of bifurcations, and which branches are recognized at various levels is decided arbitrarily. If the common ancestry of gibbons, apes, and humans is accepted, the paleontologist George Gaylord Simpson memorably wrote, "it does not much matter what rank is given to that unit."[33] During the past thirty years, many systematists have argued that taxonomy should reduce or even eliminate the use of ranks altogether. The PhyloCode, the most prominent recent proposal for taxonomic reform, in part replaces the Linnaean ranks with groups defined explicitly as branches (clades) of the tree.[34] An important theoretical justification for unranked categories is in fact a point made by Darwin in Chapter 6: breaks in the series of continuity among organisms "depend merely on the number of related forms which have become extinct."[35] Groups in a tree form a natural hierarchy, but the lengths of stems and numbers of species on branches depend upon extinction. The assignment of ranks to those branches is a matter of human preferences, not a process that emerges from the relationships themselves.

The taxonomic rank of humans remains a matter of contention even today. The current classification, in which humans and our fossil relatives rank at the tribe level, represents a major downgrade in status from the family-level rank that systematists recognized throughout most of the twentieth century. Just twenty years ago, most sources referred to humans and their fossil relatives as the family Hominidae. This set them apart from the great apes, which were classified as their own biological family: once Simiidae, then later Pongidae—a shift resulting from the suppression of Linnaeus's genus name *Simia* (the basis of Simiidae) by the International Commission on Zoological Nomenclature. Biologists could defend this distinction between hominid and pongid as compatible with the primate tree only if orangutans and the African apes were close relatives to the exclusion of humans. That idea was doubtful even in Darwin's time—he, after all, described African apes as closer to humans, and orangs as separated by a "break" from their closest allies. Many later scientists argued the anatomical case for linking gorillas, chimpanzees, and humans, and by 1960, these anatomical arguments were supported by compelling evidence of protein similarities in African apes and humans.[36] Further protein evidence during the 1960s and 1970s made it clear that humans and the African apes were very close relatives, far lower in diversity compared to most families of mammals. Yet it would take another forty years before the scientific consensus demoted humans to the tribe rank. After the notable advocacy of biologist Colin Groves,[37] the tribe-level Hominini was gradually adopted by specialists to refer to humans and their fossil relatives, but progress was slow. In 2000, the word "hominid" occurred in books a hundred times more often than "hominin." Even today some scientists hold out for a family-level rank for humans.

The idea that humans deserve a rank equivalent to the family rank of other mammals is not new. Blumenbach and Richard Owen deployed similar arguments to support their ideas that humans should be a distinct order or subclass. In the nineteenth century, the idea was that humans are separated from other animals by many adaptive features, especially brains, cooperative cultural and social systems, and upright walking. Today, some anthropologists refer to the same features to argue that humans occupy a distinctive "adaptive plateau."[38] Even if we are close relatives of apes, this way of thinking holds that classification should reflect the evidence of adaptation, not just genealogy. As a perceptive taxonomist, Simpson saw the logic of grouping apes and humans together. But he objected on the grounds of practicality: a classification must not only accord with phylogeny, it must be *useful* to those who depend on it. He wrote, "There is not the slightest chance that zoologists and teachers generally, however convinced of man's consanguinity with the apes, will agree on the didactic or practical use of one family embracing both."[39]

Darwin, who argued in favor of common descent as a basis for classification, faced the same problem. Other naturalists—even friends like Lyell and Huxley who accepted common descent—ranked humans apart because such a ranking was *useful* to them. Among the metaphysical commitments underpinning pre-Darwinian classifications was a notion of symmetry between the natural world and the human world. Placing humans at a high taxonomic rank made it possible to elevate the differences among human races as equivalent or even greater than those among other primate groups. It is no accident that the order Bimana for humans was devised by Blumenbach, who is much better known for his classification of human races. Cuvier's discussion of human races in *Régne Animal* is more extensive than his presentation of New World primates and several times longer than his presentation of lemurs. As the study of anthropology developed during the early nineteenth century, specialists treated human races at even greater length. Charles Hamilton Smith expanded Cuvier's tripartite classification of races into a 450-page volume.[40] Owen, by raising humans to the subclass Archencephala, set them, with their moral and reasoning senses, apart from other mammals and at the same time made room for himself to distinguish human groups on the same grounds.[41] Haeckel, in the second edition of *Natürliche Schöpfungsgeschichte*, published in 1870, provided a graphic illustration of his views on the variation of human races and primate species, with two opposing ranks of facial profiles.

Darwin discussed his views of races and their evolution elsewhere in *Descent* (see Chapter 7 of this book). Here, it is relevant simply to note that even though he advocated for the demotion of humans to a lower taxonomic rank, Darwin did not think that the differences among human races were small. Just within Chapter 6, he anticipated a time when "the civilised races of man will almost certainly exterminate and replace throughout the world the savage races."[42] As noted earlier, he also speculated that humans had diverged from other anthropoids in the Eocene. Darwin may have borrowed this idea of extreme antiquity for humans from Alfred Russel Wallace. In 1864, Wallace had addressed the Anthropological Society of London, a group that had splintered from the Ethnological Society in part because its founders sought explanations of human variation more in terms of race history. The majority of the Anthropological Society's members would come to favor the idea that human races had separate and distinct origins, a now-discredited idea known as polygenism. Wallace's address reflected his views that the origin of humans was extremely ancient, so that the races had been diverging for an extremely long time.[43] Darwin disagreed with Wallace about the mechanism of human evolution—Wallace thought that natural selection could not affect cultural beings, while Darwin saw the potential for greater competition in social systems. But Darwin did imagine that human evolution was as ancient as Wallace supposed. What Darwin intended with his arguments was to establish classification on the grounds of genealogy, thereby refocusing ideas of human descent onto the shared features with other primates, rather than unique features found in humans alone.

Adaptive Versus Nonadaptive Traits

Biologists sometimes use the term "Darwinian evolution" as a synonym for adaptive evolution. This term recognizes the centrality of natural selection in Darwin's way of thinking. But Darwin's own view of evolution was more expansive. He saw that natural selection could fit organisms to their environment, but he also acknowledged that some of the most informative traits were those that natural selection barely touched. When natural selection has caused species to evolve convergently, their adaptive features may mislead us about their relationships. Darwin advocated for examining vestigial traits, those that have persisted even though they have no present adaptive value, as evidence of relationships.

Early nineteenth-century morphologists might have described animals in meticulous anatomical detail, yet their classifications relied upon those traits with the greatest importance to habits. Cuvier classified animals in terms of nervous-system organization; Owen classified mammals in terms of gross brain morphology; and Blumenbach classified orders of mammals in terms of locomotion. It is no coincidence that these morphological schemes relied chiefly upon traits that differ between humans and great apes. A ladder with humans on the top rung could be built only from traits that set humans apart.

In Chapter 13 of *Origin*, Darwin expounded upon the idea that the most valuable traits for classification are those with the *least* relevance to survival and mode of life, because these are least subject to change by natural selection.[44] He reflected this idea again here, in Chapter 6 of *Descent*, with particular emphasis on the human brain. Other naturalists had viewed the human brain as the main reason to set humans apart from other primates. Darwin wrote, "A classification founded on any single character or organ—even an organ so wonderfully complex and important as the brain—or on the high development of the mental faculties, is almost sure to prove unsatisfactory."[45]

Modern phylogenetic thinking still relies upon the idea that large changes in any single organ or structure should not guide classification. Today's systematists recognize that features that have evolved uniquely within a single species cannot be used to sort that species' relationships to others. Additionally, a trait that provides useful evidence about relationships in one branch may have no relevance in other branches. Within one branch, the trait may be *primitive*, shared by various species because of descent from distant ancestors shared with many other groups. Yet the same trait may be very informative about the relationships within a much larger branch, if it is newly evolved, or *derived*, within the branch. When we look at the relationships among monkeys, apes, and humans, we know that possessing nails on the fingers is primitive because more distantly related primates, like lemurs and lorises, also have nails. Having two premolars in each quadrant of the jaw is derived—this is a trait found in some members of the group but not in distant relatives like lemurs and lorises, which typically have three premolars in each quadrant. Shared derived traits, like two premolars in the apes and cercopithecoid monkeys, provide evidence of relationships. Shared primitive traits, like fingernails, do not. The lack of a tail in living apes and humans is a shared derived trait, supporting the relationships of apes and humans, but it does not provide evidence about their relationships to cercopithecoid or platyrrhine monkeys, nearly all of which have retained tails from the distant ancestors of all primates. This way of thinking addresses a practical problem with Darwin's concept of vestigial traits: the present adaptive value of some traits may be quite obvious to biologists, while others may appear "useless" but are actually quite valuable.

Useless variations have become central to our modern understanding of the tree of life because of the growing importance of DNA sequence comparisons as evidence for biological relationships. Human genomes encompass approximately 3 billion base pairs of DNA, most of these repeated twice over on twenty-three pairs of chromosomes. Other primates likewise have genomes of approximately the same size, although deletions, additions, and duplications of DNA during primate evolution have led to much variation. Across mammals more broadly, there is still more variation, and within some vertebrate lineages vast genome size differences have evolved. These long-term outcomes of evolution resulted both from adaptation and from random chance. Many features of genomes are vestigial, including the remnants of ancient gene duplications, useless mutations, genes that have lost their functions, and fragments of ancient viruses.

Many primate groups are marked by shared useless DNA. For example, approximately 10 percent of the human genome is composed of one short sequence of 300 base pairs, repeated more than a million times and interspersed with slight variations throughout our chromosomes. These short sequences, known as *Alu* elements, derive originally from a duplication of a gene for the 7SL RNA in the common ancestors of rodents and primates.[46] After this duplication, the rodents evolved short interspersed DNA fragments known as B1 elements, while primates evolved the *Alu* elements. These are true genetic markers of the living primates. Because the proliferation continued throughout primate evolution, each branch of primates has a distinctive set, many marking the common ancestors of various groups. Another example of a useless gene marks the haplorhine primates, including tarsiers, monkeys, apes, and humans. For these primates, vitamin C is an essential vitamin, meaning that it must be present in the diet, or health problems will result. Nearly all other mammals synthesize vitamin C using the enzyme L-gulonolactone oxidase and do not need it in the diet. But haplorhines must consume the vitamin or suffer the effects of scurvy, because our ancestors suffered an ancient mutation to the gene encoding the enzyme, causing it to lose its normal function. This L-gulonolactone oxidase pseudogene, or GULOP, has become one of the most noteworthy genetic markers of our common heritage with monkeys and apes.

Conclusion

It is sometimes said that Darwin demoted humans from the top of the natural order. From the pinnacle of creation, he made humans into an insignificant twig on a vast tree of life. In Darwin's era even those naturalists who were innovators or rapid converts to evolution, such as Lyell, Huxley, and Wallace, nonetheless reserved a special place for humans in their classification. Darwin synthesized the known information about nonhuman primates and concluded that humans deserve a lowly rank. Insisting upon a classification based on common descent, Darwin found that humans are barely a subfamily. He was not the first to hold this view, but his rationale would be adopted by later biologists. The challenge of making biological classification match the branches of the tree of life continues, as does the discovery of new groups of organisms and deeper knowledge of their genomes. These discoveries have continued to drive a steady demotion of the human tribe, as our relationships to the other apes become more and more resolved.

Darwin did not perceive our demotion in rank as a reduction in status. Instead, he saw it as an elevation of the status of our true lineage.

The passage that ends Chapter 6 of *Descent* is one of the most poignant in all of Darwin's work, evoking the wonder of our deepest origins.

> Thus we have given to man a pedigree of prodigious length, but not, it may be said, of noble quality. The world, it has often been remarked, appears as if it had long been preparing for the advent of man: and this, in one sense is strictly true, for he owes his birth to a long line of progenitors. If any single link in this chain had never existed, man would not have been exactly what he now is. Unless we wilfully close our eyes, we may, with our present knowledge, approximately recognise our parentage; nor need we feel ashamed of it. The most humble organism is something much higher than the inorganic dust under our feet; and no one with an unbiased mind can study any living creature, however humble, without being struck with enthusiasm at its marvellous structure and properties.[47]

7

"On the Races of Man": Race, Racism, Science, and Hope

Agustín Fuentes

It is not my intention here to describe the several so-called races of men; but to inquire what is the value of the differences between them under a classificatory point of view, and how they have originated.

CHARLES DARWIN, *THE DESCENT OF MAN*, CHAPTER 7[1]

SO BEGINS "On the Races of Man," the seventh chapter in *The Descent of Man, and Selection in Relation to Sex*, first published in 1871. Darwin's starting point is: "Races exist, but how do they differ, and how can we scientifically understand those differences?" In "On the Races of Man," Darwin accepts the existence of biological races of humans that correspond to continental categories of African, European, Asian, American, and so on. Darwin astutely notes that these races are not separate species and that they indeed share more with one another than they differ. But at the same time, he does present and promote nonscientific and racist "evidence" to argue for their differentiation, distribution, and differential capacities. In Chapter 7, Darwin demonstrates his own substantive bias against Africans and African-descendant populations and ignores his own scientific insight, promulgating a damaging ideology about human diversity that has steadily resisted change, despite scientific refutation, for the past 150 years. While making up only one chapter of many, Chapter 7 presents what are possibly the least scientific and the most harmful assertions about human biology and behavior Darwin ever made. However, in between the biased assertions, critical aspects of Darwin's brilliance shine through, sowing seeds of hope that offer to upturn the ignorance, but only if we heed them and the scientific insight they represent.

In this essay, I will review the key arguments Darwin made about "the races of man."* First, I'll outline what Darwin said, then consider why he said it and what he might have alternatively said, given the actual science of the time, and finally lay out what we know about these assertions today. The point is not to harangue Darwin for his ignorance but to learn from his biased argumentation, correct his mistaken assumptions about human biology and history, and appreciate the small but significant insights presented in the original Chapter 7.

One hundred and fifty years after the publication of *Descent*, much of the public, including a cluster of academics, continues to repeat and perpetuate the same mistakes Darwin did. That must stop. We must challenge and refute racist pseudoscience. The best strategy is to succeed in doing what Darwin himself always wanted (and tried) to do: be a good scientist. We need to examine the variation in the world around us, construct hypotheses to explain it, and accumulate the most, and best, data possible to assess those hypotheses. And so, we shall: the second part of this chapter summarizes the science situating race and racism today, exploring what has improved, worsened, and failed to change in peoples' thoughts "on the races of man" since the initial publication of *Descent* in 1871.

---

* Darwin used the word "man," as did almost all the science writing of his time and much of ours today, to refer to humans. This is gender bias that is poor practice and unnecessarily creates a suite of unscientific implications and assumptions when discussing humanity. I repeat it in this chapter not because it is the best way to refer to humans but rather to emphasize the depth and pervasive nature of bias in the "science" (past and present).

I end this chapter with a tough but important question and a beacon of hope. The question is, "Was Darwin racist?" I'll answer that and illustrate why such a question is very important to our understanding of the science of 150 years ago and of today. I'll conclude the final section with an acknowledgment of what Darwin wished to do and what he actually got right, and why that offers hope as we move forward in the twenty-first century.

What Did Darwin Say?

I use the key section headings from Chapter 7 in the first edition of *Descent* to provide a framework for a brief summary of what Darwin said, what we know now, and what that means.

"Arguments in Favour of, and Opposed to, Ranking the So-called Races of Man as Distinct Species"

One of the main debates at the time that Darwin wrote *Descent* was whether or not the so-called "races" of humans were derived from different ancestors (the belief of polygenists) or whether they shared a distant common ancestor (monogenists).* Darwin was a stickler for clear and comprehensive presentations of information—after all, he did take eight years to write a four-volume natural history of barnacles. So, for Darwin, a critical component to set the stage for this chapter was clarifying whether or not "races" can be considered different enough to be identified as separate species and thus have distinct, and distant, origins. Darwin opens the chapter by arguing that we can look at the "races of man, viewing him in the same spirit as a naturalist would any other animal."[2] This is a classic scenario presented by many scientists who study other organisms and transition to talking about humans; it seeks to imply that one must be as "objective" when studying humans as one would be if considering a ground squirrel or an ant. Over the past

---

* Very few European or American scientists were arguing that the differences between "races" were extremely recent and/or truly negligible.

century and a half, many scholars have invoked the perspective of a scientist from another planet as narrator, assuming that such a "view from outside" keeps the science neutral and unbiased. However, in reality, this is a poor move. We know that humans are enmeshed, enculturated, and always shaped by their life experiences, language, and history. We are never fully objective when talking about humanity, even when we try to be, and a good scientist recognizes that. This was Darwin's first mistake.

Darwin does run through the case for differences and similarities in a clear, and in his view, balanced, manner. He identifies a number of superficial differences between groups of humans (the "races") and concludes that the majority of those distinctions are likely due to the fact that diverse groups of humans live in geographically distant places and have been separate for long times, so their histories shaped them in distinctive manners. To his credit, later in Chapter 7 he does suggest that such simple assumptions do not appear to hold as well for humans as they do for other animals. However, right up front in the chapter he reveals his bias, stating, "Even the most distinct races of man, with the exception of certain negro tribes, are much more like each other in form than would at first be supposed."[3] Darwin demonstrates a consistent bias against people from sub-Saharan Africa and those of African descent—a point we'll come back to.

To his credit, Darwin's review of the differences and similarities across human groups comes down strongly on the side of similarities, with a few exceptions. He spends a number of pages highlighting specific differences in bodies and behaviors and even claims that different species of lice infect different humans. This is incorrect. There are three kinds of human-specific lice, and they infect all humans. In his overview of differences (and similarities), Darwin draws from published books and studies but relies heavily on individual accounts of personal experience (including his own). This reliance on individual accounts poses a problem, given the substantive bias shown by the individuals on whom he relies (European colonialists, scientists, and travelers). However, for Darwin, the fact that the "races" can and do effectively interbreed and that they seem to grade into one another without having distinct boundaries makes them the same species. He argues that the scholars of his time cannot agree on how many "races" there are (noting sources for designating between two and sixty-three) and that bodies and morphologies show much more similarity than difference; thus he is forced to come down on the side of human "races" not being different species. In fact, he states that an understanding of the processes of evolution clearly demonstrates that the polygenists are wrong and that all humans share a distant common ancestor. As if to drive home these points, Darwin titles two sections of his chapter "Numerous Points of Resemblance in Body and Mind between the Most Distinct Races of Man" and "Each Race Not Descended from a Single Pair."

Given his overview of the "data," Darwin concludes it is most appropriate to see the "races" as subspecies. Subspecies are clusters of groups that, while in the same species (same common ancestor), have important and evolutionarily derived differences that set them apart from one another. This view, unfortunately, is exactly the same argument co-opted by racists and separatists today and remains incorrect (see "Contemporary Understandings of Human Biological Variation," below, for details).

But Chapter 7 goes beyond noting subspecific classifications. Darwin tells us "Their [the "races"] mental characteristics are likewise very distinct; chiefly as it would appear in their emotional, but partly in their intellectual, faculties. Everyone who has had the opportunity of comparison, must have been struck with the contrast between the taciturn, even morose, aborigines of S. America and the lighthearted, talkative negroes."[4] So, while overall similarities dominate the initial discussion, Darwin asserts that mental abilities are key differences and the most different (and in his view the most deficient) are people from sub-Saharan Africa and those of African descent. For example, he states, "It can hardly be considered as an anomaly that the Negro differs more," and asserts that "mulatto" women are characterized by their "profligacy," and that "hottentot"* women "offer peculiarities, more strongly marked

---

* This is a racist term developed by Dutch and British colonialists to describe the South African non-Bantu indigenous population, which is correctly referred to as Khoisan (Khoe-Sān).

than those occurring in any other race."[5] Darwin's ethnocentric, Euro-centric, and anti-African biases come through loud and clear through-out, despite his attempt at a neutral "science," looking at *Homo sapiens* in the same spirit as a naturalist would look at other organisms.

And yet, when Darwin reflects on all he's covered—even the differ-ences in intellect—he digs deep into his own experience and reveals that he is, maybe subconsciously, battling with the incongruence of what he presents as scientific "fact" and what he himself has experi-enced. He reflects:

> The American aborigines, Negroes and Europeans differ as much from each other in mind as any three races that can be named; yet I was incessantly struck, whilst living with the Fuegians* on board the "Beagle," with the many little traits of character, shewing how similar their minds were to ours; and so it was with a full-blooded negro† with whom I happened once to be intimate.[6]

He sees as a scientist, even if briefly, what he cannot see as a Briton immersed in structures and histories of European racism and bias: that the differences between people might not be what the "science" of the time states that they are.

"The Formation of Races"

"Although the existing races of man differ in many respects,"‡ Darwin writes, "as in colour, hair, shape of skull, proportions of the body, &c., yet if their whole organisation be taken into consideration they are found to resemble each other closely in a multitude of points."[7]

---

* Indigenous peoples from the southern tip of South America.

† It is not clear, but Darwin may be referring to John Edmonstone, a freed enslaved man from northern South America who taught students in Edinburgh, including Darwin, taxidermy and about the forests, animals, and ecologies of South America (Figure 7.1). Some suggest that Edmonstone was influential in interesting Darwin in embarking on travels to South America.

‡ In Darwin's chapter 7, this section ("The Formation of Races") comes after the following section ("The Extinction of Races"), but the arguments proposed and the critique I offer here flow better if we reverse that order for this discussion.

FIGURE 7.1. Reconstruction of scene of a young Charles Darwin with John Edmonstone, who taught Darwin taxidermy. (© C.C. Lucas)

Darwin was a monogenist, seeing that the evidence placed all the "races" of humans into one species. Yet, as an evolutionary scientist, his interest naturally turned to developing a suite of explanations for how one species develops into a range of similar but distinctive lineages (the "races").

The general argument for monogenists during Darwin's time was that the lineage of early humans had spread far and wide since its origin. As this spread occurred, different groups of humans became isolated from one another, more or less on the various continents, and underwent distinctive change over time (evolution) that led to their current differences (the "races"). This argument, that human "races" are the product of long-term isolation and different patterns of evolution in the areas we now call Africa, Europe, and Asia is the dominant argument that racist scholars have been using since before Darwin published *Descent* in 1871 and up through the current day.[8] It is also incorrect (see "Contemporary Understanding of Human Biological Variation," below).

There is, however, a general validity to the explanation that the widespread geographic distribution in humans has led to the broad range of human biodiversity. That is, human bodies are, in part, so variable because of our species' very wide distribution across the planet, with its diverse ecologies and landscapes. Different pressures shaped different bodies over time. This is evolution. However, this variation is not distributed in racial patterns. We know that skin color, hair type, facial features, and body shape vary quite a bit across our species but not in any pattern that clusters into continental groups (e.g., African, European, Asian, etc.; see "Contemporary Understanding of Human Biological Variation"). Dark skin, for example, occurs in distinct populations in Africa, South Asia, Southeast Asia, Micronesia, and South America. Skin color is not a characteristic that pinpoints a person to a specific geographic place of origin.

In fact, being a meticulous scientist, Darwin noticed this complex pattern of skin color distribution and understood that it countered the general notion of local ecologies shaping the characteristics of the different cases. He states,

> Of all the differences between the races of man, the colour of the skin is the most conspicuous and one of the best marked. Differences of this kind, it was formerly thought, could be accounted for by long exposure under different climates. . . . The view has been rejected chiefly because the distribution of the variously coloured races, most of whom must have long inhabited their present homes, does not coincide with corresponding differences of climate.[9]

This is right; if one is trying to use skin color to define races, it cannot be done. However, if one rejects the racial classification and just asks, "Is there a correlation between skin color and certain very broad geographic patterns?" then the answer is clear: yes. Skin color varies in accordance with the relationships between latitude, UV light intensity, and many other factors. To his credit, as a scientist, Darwin noticed this. To his fault, as a biased Briton and European, he did not follow it to the logical conclusion: skin color cannot be used to differentiate humans into racial groups.

"If, however," Darwin realized, "we look to the races of man, as distributed over the world, we must infer that their characteristic differences cannot be accounted for by the direct action of different conditions of life, even after exposure to them for an enormous period of time."[10] He also noted, "Nor can the differences between the races of man be accounted for, except to a quite insignificant degree, by the inherited effects of the increased or decreased use of parts."[11] And he even went so far as to state, "Not one of the external differences between the races of man are of any direct or special service to him."[12] In this section of the chapter, Darwin clearly acknowledges that none of the overt physical characteristics used to separate humans into "races" has an obvious or identifiable biological or evolutionary explanation supporting a racial classification. But then he reveals his bias by suggesting: "The intellectual and moral or social faculties must of course be excepted from this remark."[13] Darwin remained committed to the premise that humans are divided into significantly distinct "races"—even when his own scientific analyses suggests otherwise.

So then, how does Darwin conclude that the human "races" formed? "We have thus far been baffled," he admits, "in all our attempts to account for the differences between the races of man; but there remains one important agency, namely Sexual Selection, which appears to have acted as powerfully on man, as on many other animals."[14] However, he goes on to clarify that sexual selection likely cannot explain all racial differences,* but he remains committed to the unsupported assumption that biological "races" exist in humans. This section of Chapter 7 is the clearest example of how the power of bias can conceal what is right in front of us.

---

* Darwin also believed that because "races" could mix, their mixing might produce new "races." He states "But as with our domesticated animals, a crossed breed can certainly, in the course of a few generations, be fixed and made uniform by careful selection, we may infer that the free and prolonged intercrossing during many generations of a heterogeneous mixture would supply the place of selection, and overcome any tendency to reversion, so that a crossed race would ultimately become homogeneous, though it might not partake in an equal degree of the characters of the two parent-races" (*Descent*, 1: 241).

"The Extinction of Races"

Darwin opens up this key section of the chapter by telling us: "The partial and complete extinction of many races and sub-races of man are historically known events."[15] Here he is referring to populations and communities of people who usually, on European contact, or shortly after, diminish radically in numbers or are completely wiped out. Darwin has already identified that humans (of all "races") are amazingly capable of living in the most challenging of environments, so from his perspective, as a naturalist, the fact that many groups have gone extinct offers a quandary. He suggests a solution to this quandary by asserting, as fact, "Extinction follows chiefly from the competition of tribe with tribe, and race with race."[16] He argues that this general pattern is very old and has been characteristic of the human lineage. However, his analysis is inspired by the specific case of European expansion, which he explains by telling the reader, "When civilised nations come into contact with barbarians the struggle is short, except where a deadly climate gives its aid to the native race."[17] He adds: "The grade of civilisation seems a most important element in the success of nations which come in competition."[18] He sees this outcome (group or "racial" extinction) as the result of direct competition: a group or "race" wins because it is more "civilized." He also assumes that when "civilized" (meaning European) groups change the landscape, it "will be fatal in many ways to savages, for they cannot, or will not, change their habits."[19] He acknowledges that diseases can, and do, play a role and that "the evil effects from spirituous liquors, as well as with the unconquerably strong taste for them shewn by so many savages,"[20] also has deleterious impact.

Here Darwin is asserting the specific (and still common) belief that the genocide of indigenous peoples at the hands of European colonizers is due to:

a) the native peoples being naturally outcompeted by more "civilized" groups, and

b) weakness on the part of the indigenous peoples—of mind, of constitution, of an inability to forgo spirituous liquors, of an inability to adapt to the "civilized" lifestyle.

In the first edition of *Descent* (1871), Darwin spends little space on this section, offering a summary of his thoughts on the matter rather than a more comprehensive review of information. However, it is extremely relevant in this one instance to mention what Darwin added to this section of Chapter 7 in the second edition of *Descent*, published a few years later, in 1874, which is the version most commonly reprinted and read. He lengthened the section and offered multiple examples in support of his assertions, more expansively arguing his case. He also crystalized his erroneous, but powerful, "natural" argument for genocide and colonialism.

In the second edition, a chunk of this section outlines "case studies" of the post-European-contact devastation of the Tasmanians, the Maori of New Zealand,* the New Hebrideans (today Vanuatuans), the Andaman Islanders, and others. In these pages, as Darwin describes the crashes and extinctions of populations, he also clearly documents the horrors and atrocities of European colonial contact. He describes massive stress leading to widespread infertility and infant mortality. He identifies the introduction and impact of infectious diseases, acknowledges displacement and forced movements, and suggests the inability of the indigenous populations to adapt. Darwin sets up these genocides to be seen as outcomes that ensue due to natural selection, the natural outcome of competition between "races." He argues that the entire process can be compared to the functioning of systems of the "lower" animals, maintaining that certain animals do better, are better able to adapt to challenging circumstances, and possess higher levels of health and vigor than others (citing his earlier work in that vein). He compares indigenous populations to certain lower animals, suggesting that "savages" are likely to respond poorly when challenged with a sudden change of lifeways. He claims, "Civilised races can certainly resist changes of all kinds far better than savages."[21] Here Darwin comes very

---

* He contrasts the Maori with the Irish, who he states underwent very similar scenarios but grew in population size rather than diminishing—implying that the Irish are a more hardy "race."

close to asserting that the genocidal effects of expansion and colonialism are the logical and expected outcomes of natural laws.

I'll end the review of this section with Darwin's own words in the last paragraph in the first edition of *Descent*, where he clearly summarizes his thoughts.

> Although the gradual decrease and final extinction of the races of man is an obscure problem, we can see that it depends on many causes, differing in different places and at different times. It is the same difficult problem as that presented by the extinction of one of the higher animals—of the fossil horse, for instance, which disappeared from South America, soon afterwards to be replaced, within the same districts, by countless troops of the Spanish horse. The New Zealander seems conscious of this parallelism, for he compares his future fate with that of the native rat almost exterminated by the European rat. The difficulty, though great to our imagination, and really great if we wish to ascertain the precise causes, ought not to be so to our reason, as long as we keep steadily in mind that the increase of each species and each race is constantly hindered by various checks; so that if any new check, or cause of destruction, even a slight one, be superadded, the race will surely decrease in number; and as it has everywhere been observed that savages are much opposed to any change of habits, by which means injurious checks could be counterbalanced, decreasing numbers will sooner or later lead to extinction; the end, in most cases, being promptly determined by the inroads of increasing and conquering tribes.[22]

Contemporary Understanding of Human Biological Variation, in a Nutshell

In order to best assess the lasting impacts, positive and negative, from Darwin's Chapter 7 on the "races" of humanity, we need to briefly review what we know in 2021, 150 years after the publication of the first edition of *Descent*. In other words, what data and understandings would Darwin be working with if he wrote Chapter 7 in 2021, as opposed to 1871?

Subspecies?

Substantive evidence from paleoanthropology, human biology, biological anthropology, and genomics demonstrate clearly and unequivocally that all contemporary humans belong to one species and one subspecies: *Homo sapiens sapiens*.[23] Across the evolution of the genus *Homo* over the past ~2.5 million years, there were multiple lineages, possibly different species, and different subspecies (e.g., Neanderthals, Denisovans, *Homo erectus*, *Homo naledi*, etc.).[24] However, for at least the past ~35,000 years, there has been only one kind of humans on this planet, us—*Homo sapiens sapiens*.

Biological Races?

> Race does not provide an accurate representation of human biological variation. It was never accurate in the past, and it remains inaccurate when referencing contemporary human populations. Humans are not divided biologically into distinct continental types or racial genetic clusters. Instead, the Western concept of race must be understood as a classification system that emerged from, and in support of, European colonialism, oppression, and discrimination. It thus does not have its roots in biological reality, but in policies of discrimination.
>
> AMERICAN ASSOCIATION OF BIOLOGICAL ANTHROPOLOGISTS (AABA) STATEMENT ON RACE AND RACISM[25]

There is no evidence to support the division of contemporary humans into specific biologically defined clusters that correspond to "race" categories. The American Society of Human Genetics notes that "genetics demonstrates that humans cannot be divided into biologically distinct subcategories"; that "the study of human genetics challenges the traditional concept of different races of humans as biologically separate and distinct"; and that "genetics exposes the concept of 'racial purity' as scientifically meaningless."[26] This same content is reiterated in similar forms in the position statements of the American Anthropological

Association, the American Association of Biological Anthropologists, and many other scientific associations that focus on human evolution, human biology, and human behavior.

All humans are identical across more than 99 percent of their genome, and the < 1 percent of genetic variation in our species is widely distributed. Thousands of populations of humans across the planet can be differentiated by clusters of patterns of genetic variation, but none of these variations are evolutionarily exclusive and none of the clusters define any populations as discrete enough to be considered a distinct biological lineage (a subspecies). The distribution of human DNA sequence variation does not map continental groupings, such as "African," "Asian," and "European," as distinct, relative to other possible groupings. Movement, migration, and population mixing are characteristic of much of human history, and our genome diversity and distribution of genetic variation demonstrates that.[27]

Human morphology (how we look) varies widely, and humans are among the most diverse mammals in body shape and size. Human skin color varies from dark to light; human hair color and types vary substantially. Human populations show a range in sizes, with average heights varying from over 1.8 meters to under 1.5 meters and weights from nearly ninety kilograms to under forty-five. More than twenty-eight human blood groups vary across the planet, as do multiple aspects of human enzyme activity, disease resistance and susceptibility, and other characteristics of physiological function. Human bodies vary enormously, but none of that variation is distributed into specific clusters that match the concept of race (e.g., African, Asian, or European). Tall and short, heavy and light populations are found across all the major areas of the planet, not confined to one geographic locale or continental point of origin. Skin color varies across the planet, with darker skin characterizing populations across much of Africa, Asia, and the Americas, and lighter skin appearing in more northerly areas of Eurasia. Similar patterns hold for all other types of morphology. Our physical appearance and our physiologies vary enormously, but they do not divide humans into biological races or continental types.[28]

While "race" is not a biological category, race as a social reality—as a way of seeing people, structuring societies, and experiencing the world—is very real. Societies construct racial classifications not as units of biology but as ways to lump together groups of people with varying historical, linguistic, ethnic, religious, and other backgrounds. These categories are not static. They change over time as societies grow and diversify and alter their social, political, and historical makeups.* The American and European histories of creating social races and of structuring their societies around racial inequalities are well documented.[29] These processes and patterns are deeply rooted in the assumptions of "natural" differences between Europeans and those people from all other places on the globe. The differences evident in Darwin's biases (and in those of so many people today) are not present in the actual data of human variation. These systems of racial classification are tied to histories of expansion and contact, of colonialism, empire, and slavery. They are rooted in classifications of different human beings as systems of justification for exploitation and oppression.[30]

However, while "race" is not biology, racism† can certainly affect our biology, especially our health and well-being. Substantial research demonstrates that racialized social structures, from overt oppression and physical subjugation to access to health care to economic and educational discrimination to histories of segregation and material deprivation to one's own racialized self-image as a result of such systems, can impact the ways our bodies, immune systems, and even our cognitive processes react and develop.[31] This means that "race," while not a biological division, can have important biological implications because of the effects of racism. The belief in "races" as natural divisions of human biology and the structures of inequality (racism) that emerge from such beliefs are among the most damaging elements of the human experience both today and 150 years ago.

---

* Consider Darwin's viewing of the Irish as a race and how they are viewed today

† "Racism" is any prejudice against someone because of their race, when those views are reinforced by systems of power. See Ijeoma Oluo, *So You Want to Talk about Race* (New York: Seal Press, 2018).

What Has Gotten Better, Worse, and Not Changed since Darwin Wrote about "Race"?

Darwin grew up with, and was educated into, the belief that humans are divided into biological "races" and that these "races" are ranked from the lower "savages" (most of the world) to the higher "civilized" Europeans. He believed that the "race" called "African" was at the bottom of the primitive-to-civilized hierarchy. He believed that there were evolutionary (biological) reasons for the existence of differences between the "races." He was wrong on all counts. Unfortunately, some of these beliefs are still present in society today, with only slight modifications.

While Darwin's attempt to explain the origins of "races" was unsuccessful (as he himself noted), he did propose that sexual selection (differential mating pressures) had an influence in forming the races. This has been, and still is, used in some contemporary racist and nationalist (separatist) thought about miscegenation (race-mixing) and its threat to "racial purity."* This interpretation is incorrect but remains a strong and lasting racist myth.

On a positive note, our scientific understanding of genetics, human biology, and human evolution has advanced substantially since Darwin's time. We can clearly and explicitly refute the hypothesis that humans are divided into "races" (e.g., African, European, Asian, etc.) that are biological units. It is unfortunate that a substantial segment of the public does not understand or accept this and even more grievous that some scholars ignore the scientific data and continue to make racist assertions as if they were valid.[32]

Given what Darwin wrote and his acumen as a scientist, I have no doubt that if writing his chapter "On the Races of Man" today, Darwin would champion the lack of biological "races" and spend a great deal of time reviewing the data for human variation in order to demonstrate that fact. When reading the original Chapter 7 closely, it is obvious that even in his time the actual data refuted much of the argument for

---

* It was only in 1967 that anti-miscegenation laws were ruled unconstitutional in the United States.

biological "races." In the pages, you can see Darwin struggling to make sense of that (for the most part). One of his main conclusions, almost hidden in a single sentence at the end, is: "We have thus far been baffled in all our attempts to account for the differences between the races of man."[33]

Was Darwin Racist?

The neurobiologist Steven Rose wrote, in an essay entitled "Darwin, Race, and Gender":

> Darwin was, after all, a man of his time, class and society. True, he was committed to a monogenic, rather than the prevailing polygenic, view of human origins, but he still divided humanity into distinct races according to differences in skin, eye or hair colour. He was also convinced that evolution was progressive, and that the white races— especially the Europeans—were evolutionarily more advanced than the black races, thus establishing race differences and a racial hierarchy.[34]

It is well documented that Darwin was an abolitionist and saw slavery and general race-based cruelty as horrific and unjust.[35] However, if "racism" is any prejudice against someone because of his or her race, when those views are reinforced by systems of power,* then yes, Darwin was racist. His overt bias in regard to the mental, moral, and social capacities of humans from the continent of Africa, Afro-descendant populations, and indigenous peoples of the Americas was clear in *Descent of Man* and other writings. Darwin's racism was neither intentional nor malicious, but it is an example of how racism is maintained—not by the vitriolic screaming and overt acts of violence by a minority but rather by passive acceptance of a particular "reality" and promulgation of the status quo by a majority. Participation in this pattern of racism was Darwin's greatest failure as a scientist and the singular missed chance for good with Chapter 7 when it was originally published. It's clear that Darwin saw

---

* Which it is. See Oluo, *So You Want to Talk about Race.*

that the "race" ideology he accepted and endeavored to explain did not fit with the available data or his own life experiences. Yet he stuck to it. One does not have to harbor malice for one's racism to have truly malevolent and significantly damaging effects. If one has respect and prominence, then the damage is done. And Darwin had both. His words in Chapter 7 acted to bolster racist (and false) ideologies. To this day racists and nationalist/separatist ideologues use Darwin's words and general arguments as basis for their erroneous and intentionally hurtful and hateful positions and actions.

Darwin was, like much of humanity, a biased human being who's at least a little bit racist. So many humans are that way because the societies that raised them are deeply structured with racist, classist, and gendered divisions central to their histories and contemporary functioning. Life experiences in such environments shape minds and bodies.[36] And like Darwin, most of us often sense, even if subconsciously, that the realities that racist structures try to force us all to see are not an accurate description for the diversity of the human species. Nor do such racist structures represent the only way we can live together.

Careful analyses and thorough investigation, even when their results do not sit well with one's preconceived notions of how the world "is," move science forward. It is in the close reading of Chapter 7 that we can see what Darwin got right as a good scientist and why that matters and offers hope. His scientific eye caught the inconsistencies and misrepresentations in the assertion that "races" are biological divisions of humanity, even if his general worldview did not change to reflect this insight. Today, in 2021, with access to the contemporary data on human biological diversity and evolutionary history, Darwin would not have concluded, "We have thus far been baffled in all our attempts to account for the differences between the races of man." He would have affirmed, conclusively, that the division of humans into biological races has no support from the biological and evolutionary sciences.

8

Resolving the Problem
of Sexual Beauty

Michael J. Ryan

The sight of a feather in a peacock's tail, whenever I gaze at it, makes
me sick!

CHARLES DARWIN, LETTER TO ASA GRAY, 1860[1]

CHARLES DARWIN PENNED THIS COMPLAINT, in a letter to the
famed North American botanist Asa Gray, in 1860. Darwin was physically sick much of his adult life, but this ailment was psychological, most
likely a result of cognitive dissonance.[2] The resolution of this revulsion
toward the peacock's tail serves as a compelling metaphor for Darwin's
theory of sexual selection.

The peacock's tail is a stunning ornament and somewhat of an icon
for the sexual beauty that surrounds us in nature (Figure 8.1). Melodious
songs of birds, sparkling colors of fishes and butterflies, overpowering
choruses of frogs and crickets, the musky odor of deer, and all the cultural accouterments of courtship in our own species are only some examples. Darwin had a profound understanding and appreciation of the
natural world, including its ubiquitous sexual beauty. So why did this
particular form of beauty bother him so much?

FIGURE 8.1. A male peacock displaying his tail to a female peahen. (iStock by Getty)

This work is celebrating the 150th anniversary of *The Descent of Man*, but that is not the full title of the work, which reads: *The Descent of Man, and Selection in Relation to Sex*. Part 2 of the book lays out the theory of sexual selection and then surveys several animal groups, to which Darwin applies this theory to explain the evolution of traits akin to the peacock's tail. In the final chapters of the book, he applies sexual selection to humans, especially asking how sexual selection might be important in generating population variation in morphology and behavior within our own species. The purpose of this chapter is to give an overview of sexual selection, Darwin's second great theory.

Natural Selection Is Not Sexual Selection, or at Least It Wasn't to Darwin

In order to understand sexual selection and all the fuss that surrounded it, we first need to have a clear understanding of natural selection, the main topic of Darwin's 1859 book, *On the Origin of Species*.

Darwin's main premise of that book can be summed up in three words, "descent with modification." He argued that all organisms at some point in their history share a common ancestor, and when they diverged from that ancestor, over time, many of their traits were modified to enhance their survivorship. Take the finches Darwin discovered on the Galápagos Islands as an example.[3] They all shared a common ancestor, a finch species of the mainland of Central or South America that first arrived on the Galápagos Islands about 2.3 million years ago. The descendants of these migrant finches subsequently invaded the different islands of the archipelago, where they diverged from one another in a number of characteristics. Most notably, beaks evolved in response to natural selection to exploit different local food sources. Beaks can vary in size and shape, and there are specific genes responsible for this variation.[4] Selection will favor those variants that enhance foraging performance in the local environment. On some islands, selection favored large beaks that were good for cracking seeds and nuts; on other islands, selection favored long, pointed beaks that allowed the birds to extract insects from their hiding places; and in still other cases, selection favored the evolution of medium-size beaks that were adapted for catching insects on the wing (Figure 8.2). As the birds diverged from one another in morphology, they also diverged in their behavior, their songs, and their genes. These differences, in combination with the fact that the populations on different islands were separated by ocean, inhibited mating between the populations and ultimately resulted in populations becoming different species.

Charles Darwin and Alfred Russel Wallace were codiscoverers of the theory of evolution by natural selection. Their joint paper was read at a meeting of the Linnean Society of London on July 1, 1858, and was published in the society's journal later that year. A year later Darwin published a more detailed account of this theory in *On the Origin of Species*.[5] Despite the book's title, his main emphasis was on the role of natural selection in the evolution of adaptations.

There are three factors that need to conspire for an adaptation to evolve. The first is an insight that Darwin gleaned from his readings of Thomas Malthus's *An Essay on the Principle of Population*: the rate of

FIGURE 8.2. Diversity of beaks in Galápagos finches. (© National Geographic)

reproduction of the population will often far exceed the resources in nature that can support it.[6] Thus, not all of the individuals that are born into a population survive to reproduce. The second part of this theory is that survival is not random. Certain individuals survive because they have traits that enhance their survivorship. These are the traits that are favored by natural selection. But selection does not lead to evolution unless those favored traits have a genetic component. This is the third piece of this puzzle. When traits favored by natural selection can be inherited, genes associated with these traits increase in subsequent generations, in turn causing an increase in the frequency of these traits that enhance survivorship. Of course, Darwin did not know about genes; Mendelian genetics (the principles of inheritance Gregor Mendel formulated in 1865) was a few years off, but the notion of heredity was central to animal and plant breeding at the time and well known to Darwin and most other scientists. This third piece is how adaptations evolve; it is why Galápagos finches have those exquisite beaks that are so good at what they do.

The power of the theory of natural selection is in its brilliant simplicity and the fact that it generates predictions that have been scientifically verified over the ages. Evolution by natural selection is one of the most well-established facts in all of science.[7]

In 1860, Darwin was well recognized for explaining how organisms' adaptations evolved for survival. But when he gazed at the peacock's tail, it stared back at him as a stark challenge to this theory of natural selection. It was obvious, at least to Darwin, if not to Wallace, that the peacock's tail was not an adaptation for survival, more likely hindering survival than promoting it. As the male is signaling his presence to potential mates, he is also being eyed as a potential meal by a variety of predators from tigers to mongooses. His tail makes him more attractive as a mate but more conspicuous as a meal. If the peacock's tail was a single aberrant glitch in the workings of natural selection it might not have been such a focus of Darwin's consternation. Darwin acknowledged, however, that these types of traits, those that seem maladaptive for survival, are ubiquitous in the animal kingdom. Long tails, bright colors, elegant courtship dances, and elaborate vocalizations all seem to invite death, not to circumvent it. How could this be explained?

There were some immediate hints. These types of traits were usually more developed in males than in females; in fact, they were often absent in females. Peahens (female peafowl) have tails, but they are dwarfed by the size of the males' tails. Also, males usually employed these traits in the breeding season, and many of these traits were part of the animal's courtship display.

Darwin's theory of sexual selection was his explanation for the peacock's tail and other courtship traits that were partners in the crime of seemingly reducing survivorship. This theory followed a three-step road map, similar to natural selection. First, unless there is perfect monogamy, there are often more males trying to mate than there are females willing to mate. Second, mating success is not random; the successful males have traits that enhance their ability to gain direct access either to females or to resources needed by females, or the males have traits that enhance their beauty, making them more attractive as mates. Third, these traits have a genetic component, resulting in traits that enhance mating success being disproportionally promulgated in subsequent generations.

It was not difficult to resolve the conflict between natural selection and sexual selection. Traits evolve only if they are passed on to the next generation through mating, but animals can mate only if they survive. Thus, a male peacock with an abnormally short tail might survive quite well but never mate, while a male with an abnormally long tail might be quite attractive to females but likely would not live long enough to try out his tail. Thus, in many cases, there is a conflict between natural selection and sexual selection, and the traits that evolve strike some balance between survivorship and mating success.

Sexual Selection Is Borne out of Sexual Conflict

Why is there a sex bias in sexually selected traits? Why is it typically the males that have to compete for females and not vice versa? Of course, there is a lot of variation in mating systems; in seahorses and pipefish the sex roles are reversed, and females compete for males; and in a larger number of species, such as sticklebacks and humans, there is mutual mate choice. But why are these mating systems the exception rather than the rule?

The answer is as fundamental as can be imagined. Many of the differences between males and females, in almost all the species known to reproduce sexually, result from differences in gamete size. In humans and in most other animals, a female's eggs are the largest cells in her body, while a male's sperm are the smallest cells in his body. Females invest much more in their gametes than do males, even though males produce many more gametes. In humans, females have a few hundred eggs, while a male will produce a few billion sperm during his life.

Gamete size is the fundamental character that defines an animal's sex. Many would think that an individual's genitals might be diagnostic of its sex, but those many would be wrong. As a case in point, there is a species of lice that lives in caves in Brazil and feeds on bat guano. This is a typical sexually reproducing species, in that there are males and there are females. But atypically, the females have "penises" and the males have "vaginas." A female inserts her penis into the male's vagina and sucks up the sperm from the inside of his body into her body, where his sperm fertilize her eggs.[8] It is not a very typical mating strategy, but these lice are quite typical in that the females have the large gametes and the males have the small gametes.

The difference in gamete size mirrors the difference in reproductive investment by the sexes. In a landmark paper, published in 1972, that revitalized interest in sexual selection theory, Robert Trivers pointed out that this difference in reproductive investment results in different mating strategies between the sexes that, in turn, results in a conflict of interests when it comes to how males and females maximize their reproductive success.[9] In shorthand, we refer to this as sexual conflict.

One offspring arises out of one sperm fertilizing one egg. Thus, there are many sperm competing for access to a few eggs, which translates to many males competing for fewer females. This simple fact has caused many evolutionary biologists to question why males even exist, a conundrum referred to as the "cost of males."[10] If females reproduced clonally, with no genetic input from males, all of their offspring, all of which would be daughters, would bear offspring. If females in a similar population reproduced sexually, only half of their offspring would be female, only their daughters but not their sons would bear offspring, and the sexual population would grow at a much slower rate than the asexual population. The short answer explaining the existence of males is that sexual reproduction provides a means for creating genetic variation among offspring, even though it slows the rate of population growth.[11]

Selection favors both males and females to produce as many offspring as possible, but the difference in the size of gametes can result in two different mating strategies for the sexes. Males are selected to mate often, while females are selected to mate carefully. A famous experiment with fruit flies produced in 1948 by Angus Bateman illustrates this dichotomy.[12] Male and female flies were allowed to mate between one and four times. The number of offspring for each fly was counted. The more often a male mated, the more offspring he produced, while mating with more males had little effect on the female's reproductive success.

Bateman's study has generated some controversy as well as some additional support. The controversy comes from the interpretation that we should always expect males to mate promiscuously and females to choose mates carefully. This controversy is especially heated when applied to our own species, sometimes being used to justify promiscuity in men while enforcing monogamy in women.[13] Genetic markers, however, have now revealed that in some species females can be promiscuous and that additional matings can influence her reproductive success. On the other hand, it does seem that in many if not most animal mating systems, the number of matings has a greater effect on male reproductive success than it does on female reproductive success, and the exceptions to this pattern actually lend support to Bateman's theory.[14] Also, in many species almost all of the females will reproduce about the same number of times, while many males will go unmated and a smaller proportion of males will garner a greater proportion of matings. Thus there is more variation in mating success among males than there is among females. Since selection acts on variation, sexual selection will be more intense on males than on females.[15]

Competition between males for females is also heightened by what is called the operational sex ratio.[16] Even if the sex ratio of a population is balanced, the number of males and females ready to mate at any point in time is often biased toward males. Again, this has to do, in part, with the differences in gamete size. Once a female's eggs are fertilized, she must nurture them to gestation if there is internal fertilization or begin to develop a new set of eggs if fertilization is external. Thus, once a female mates, she is out of the mating game for a while, but in most species males do not have the same constraints and are able to continue mating. Consider our own species: once a woman conceives, she cannot do so again for another nine months, while her mate could continue to fertilize other women almost immediately. If a woman had sexual intercourse 100 times during gestation, she would still probably birth only one child. A man with similar sexual proclivities might have sired a small village in that time. Thus, the ratio of reproductively ready males to reproductively ready females (i.e., the operational sex ratio) results in many more males competing for many fewer females. How does this competition play out?

How to Get a Mate

In general, there are two ways males can get mates. One is competition. Males might compete for direct access to females. For example, in a baboon troop, dominant males aggressively prevent copulations by lower-ranked males when females are ovulating. Males might also compete by defending resources critical to females. For example, male red deer defend areas where females congregate for grazing. The evolution of animal weapons is most often in response to sexual selection for fighting with other males to get mates.[17]

The peacock's tail and other traits of sexual beauty can hardly be considered weapons. These traits arose from another form of sexual selection—female mate choice. In these mating systems, females are in the driver's seat when it comes to deciding who gets to mate. Males evolve traits that make them more sexually attractive to females. These are the traits that constitute much of the sexual beauty that we see in nature. Most of these elaborate traits, such as the peacock's tail, evolved in the service of sex. Female mate choice based on elaborate male traits is especially important in mating systems in which males do not offer resources to females or have the opportunity to gain access to females directly. These mating systems, often referred to as lek-like, are characterized by males gathering together to advertise for females and females surveying the males and then freely exercising their choice for a mate. In sage-grouse, for example, sometimes hundreds of males congregate in a breeding ground, or lek, in the sagebrush plains of North America from March until May. Females visit the leks, often congregating in the territories of just a few males, where they select mates. It is not uncommon for fewer than 10 percent of the males to be responsible for more than 75 percent of all the matings.[18] Although true lek systems are usually restricted to birds, this type of mating system, in which females reign supreme, can occur in almost any type of animal.

Nature is diverse and especially so when it comes to the mating game. In some species females compete for males, and males choose their mates. This is often referred to as sex role reversal, and a classic example is that of seahorses and pipefish. In these species, males become "pregnant" in that they carry around the developing embryo in their pouch, during which time they usually do not mate. Once a female deposits her eggs in a male's pouch, however, she is then free from the responsibility of maternal care and can seek out other males for mating.

Another exception to the strict male competition–female choice paradigm is that of mutual choice. Our own species might be one of the best examples. In some societies, men compete for the attention of women during courtship, and women are often able to choose the men with whom they pair, but men also will choose among females as potential partners. Sometimes men and women base their choice on similar characteristics. Other times, however, they use quite different criteria; for example, some studies have proposed that men tend to weight a partner's age and physical beauty more heavily, while women are more attendant to the resources of a potential mate.[19]

There are also exceptions to the two tactics, direct competition and advertising, that males can employ to gain matings. Some mating systems are characterized by alternative mating tactics. Take freshwater sunfish as an example. Some large colorful males defend areas that are ideal for mating.[20] These fish have external fertilization, and as females release their eggs into the water column, males simultaneously release their sperm, and fertilization takes place when the gametes of the two sexes happen to make contact with one another. Some males, however, resemble females, and this allows them to loiter near these territorial males. When the territorial male and his female start to shed their gametes, the male who is impersonating a female begins to broadcast his sperm. In addition, so-called sneaker males, who are small and inconspicuous, will dart past a mating couple while broadcasting his sperm, hoping to steal a fertilization. In some cases, alternative mating strategies have a strong genetic component, such that courting males father courting sons while sneaker males father sneaker sons. In other cases, males are making the best of a bad situation, and genes might have nothing to do with these mating strategies, which instead arise from environmental influences. If a male has limited access to resources during development, for example, he might mature at a small size. A small courting male would probably be less attractive to a female than a large courting male, and consequently the small male adopts the alternative mating strategy of sneaking.

There are many exceptions to the "typical" mating systems in which males compete for females and females choose their mates. But all the exceptions prove the rule of Darwinian sexual selection theory rather than challenge it. Then why, we must ask, has sexual selection by female choice always been shrouded in controversy?

Darwin versus Wallace and the Controversy of Female Mate Choice

There were two main criticisms of Darwin's theory of sexual selection by female choice. The first cited the lack of compelling evidence that females choose mates. The second criticism was that even if it were to occur, Darwin could not explain why females have these preferences.

Darwin and Wallace did not part ways after their codiscovery of natural selection theory. They became lifelong friends and scientific confidants and mostly agreed in their interpretation of how natural selection shapes the world around us. But the same cannot be said when it came to Darwin's second great theory, sexual selection. As Wallace noted:

> Two or three considerations appear fatal to the theory [of sexual selection]. . . . In the first place, it seems quite incredible, without direct evidence on the point, that a large majority of the females of any species, over the whole area of its range and for many successive generations, should agree in being pleased by the same particular kind of variation.[21]

Wallace was not the only one of Darwin's contemporaries to reject his theory of sexual selection. For many, the idea of males competing "red in tooth and claw"[22] for access to females seemed a proper extension of the theory of natural selection. The notion, however, that females had strong preferences for males based on the details of their courtship behavior and morphology, details that seem only to hasten the male's demise, was a real stretch. The Victorian notion that male mate choice reigned supreme in our own species made it seem improbable that in other species it was the females who wielded similar powers. It is notable that, contrary to these Victorian biases, Darwin implicitly, and sometimes even explicitly, endowed female animals of many species with superior power in the mating game. But at the same time it is confounding that, like a typical Victorian, he viewed women as inferior and devoid of such power, as is clearly documented in the next chapter by Holly Dunsworth.

Wallace too was a Victorian, so how did he explain these traits of sexual beauty, such as the sexual dimorphism in plumage coloration in birds? Rejecting Darwin's idea that sexual selection favored males with bright colors, Wallace suggested that natural selection favored females with dull colors because of the advantage being camouflaged afforded them.

The second criticism of sexual selection by female choice was that Darwin did not offer a cogent explanation for why females would have such preferences for males. His suggestion that females had an aesthetic sense much like our own seemed to be kicking the can down the road. It led only to the question of why the females would have a taste for the beautiful. When Wallace did give female choice some credence, he suggested that females were assessing traits that indicated a male was vigorous not beautiful. This difference of opinion between Darwin and Wallace, which they never resolved, is wonderfully documented in a recent book by Evelleen Richards.[23] This disagreement that started between the two Victorians continues until this day.

The Evolution of Female Mate Choice

There is now no controversy about the existence of female mate choice.[24] Hundreds of studies have shown, through natural observation and experimental manipulation, that females of many taxa, including crickets, fish, frogs, and birds, show preferences for the traits that are more elaborately developed in males and play some obvious role in courtship behavior. Numerous studies show that females prefer brighter colors, faster dances, longer tails, louder calls, more complex songs, and stronger odors.

The Darwin-Wallace controversy as to why females have such preferences is sometimes so heated it is forgotten that in many cases there is no controversy. To explore some of these arguments, I will borrow from an example in my recent book *A Taste for the Beautiful: The Evolution of Attraction*.[25]

Now that it is well documented that females exhibit preferences for sexually selected traits, the next challenge is understanding why females show these preferences. There are three main processes that can bring about the evolution of mate choice. One, females can evolve preferences for traits of males who will deliver immediate benefits to the female, such as mating with the male who will fertilize more eggs. Two, traits and preferences can become genetically linked, such that male courtship traits evolve because they are favored by females, and the female preference evolves only because it is genetically linked to the traits under selection. Three, traits can evolve if they exploit hidden preferences; these are preferences harbored by the female but not expressed, as the traits that elicit them have yet to evolve. I will review these different processes in turn.

For the sake of argument, we will assume that there are genes influencing both female preferences and male traits, and that males and females contain both preference and trait genes but only females exhibit preferences and only males exhibit the sexually selected traits.

To explore these different scenarios for the evolution of female preferences we will consider the red-winged blackbird. Males perch on top of vegetation, quite often cattails, in early spring, singing as they flash bright red epaulets, or patches on their wings, in an exhaustive effort to advertise themselves to females. Why do males have these flashy colors, and why are females attracted to them?

Imagine that a closely related species, the yellow-headed blackbird, begins to nest in the same marsh as the red-winged blackbird, something that can happen in nature. As with most matings that take place between different species (i.e., heterospecifics), if a female red-winged mates with a male yellow-headed blackbird, her eggs will not be fertilized. Thus, there is a premium on females being able to recognize their own species and discriminate against other species. The easiest way for a female red-winged to identify a male of her species is through the red epaulets. Imagine there are two types of females, those that are more attracted to males with the red epaulets and those who do not discriminate and instead mate randomly with red-winged and yellow-headed males. Females with the preference for red wing patches will sire offspring, and if there is a genetic basis to their preference for red, this preference will be passed on to the next generation. Females that mate with yellow-headed blackbirds will not produce offspring, as her eggs will not develop, and her genes will never swim in the next generation's gene pool. Eventually, through time, all the female red-winged blackbirds preferentially mate with their own species over the yellow-headed blackbirds. Female preference for males of their own species is the most common form of mate choice and the best example of its utilitarian benefits.[26]

Imagine another visitor now invades the marsh, but it is not another species of bird—it is feather lice. These lice have a variety of ill effects on the birds' health, and they also influence the brightness of the birds' plumage. Females already prefer males with red epaulets, because this preference delivers mates of the same species, but now selection will also favor females that prefer brighter red epaulets, because these females will avoid contracting sexually transmitted parasites. Females who don't discriminate based on the intensity of the red coloration are more likely to become infected with lice themselves and will suffer a decrement in their reproductive success—sick females do not lay as many eggs as healthy ones. Those females who prefer both the red-winged males and the brighter red-winged males garner two types of direct benefits from their mate choice: they mate with the correct species and the healthier males of their own species. These are called direct benefits because they directly influence the number of offspring a female produces.

Females can also garner indirect benefits through mate choice. Indirect benefits influence not the number of a female's offspring but instead the quality of her offspring. We know from many studies that some animals can be more parasite-resistant than other animals, and there can be a genetic component to this parasite resistance. Females who prefer more brightly colored males will produce offspring, both sons and daughters, who will be more resistant to parasites. Selection favors the evolution of these parasite-resistance genes in birds because individuals with these genes survive longer. The female preference genes, those genes underlying the preference for redder males, now end up in offspring that also have the parasite-resistance genes. The female preference genes, in this case, increase in frequency not because they are directly favored by selection but because they are in the bodies of individuals with parasite-resistance genes that are under direct selection. This is how a female preference for a male's "good genes"—genes that enhance survivorship—can evolve. The female preference genes hitchhike a ride into the next generation along with the good genes for survivorship. Studies of peacocks and frogs document how good genes preferences operate in nature.[27]

Another case in which female preferences can evolve through genetic hitchhiking is called runaway sexual selection or the "sexy son" hypothesis. In this scenario, the males who are bright versus dull red do not differ in their survival abilities. Even though most females are mating at random relative to plumage brightness, some of the females mate only with the redder males; therefore, these males will increase in subsequent generations. Offspring of these matings will have genes for redder coloration and the preference for redder coloration. As noted above, the sons express only the trait, while the females express only the preference. The genes for brighter wings will evolve because they are favored by female choice, and the genes for this female preference will evolve through genetic hitchhiking because they are present in these more attractive males, the sexy sons. This has been a difficult hypothesis to test empirically, but studies of stalk-eyed flies bear out the major predictions of runaway sexual selection.[28] Thus, both good-genes preferences and runway sexual selection can result in the evolution of female preferences through genetic hitchhiking.

We will consider one final scenario that can bring about the evolution of elaborate courtship traits and preferences for those traits. In this case, males evolve traits to tap into hidden preferences in the female; these are "preferences" that might have evolved for other tasks, such as foraging, but then become co-opted by mate choice. Again, we will consider the blackbird example but further back in time, before males evolved the red epaulets. In this scenario there is no benefit to a mutation that causes red epaulets, but there is a cost, as predators are more likely to spot these birds. If there is all cost and no benefit, these mutations quickly go extinct, as the bearers of this mutation quickly become meals for predators instead of mates for females. Now, let's imagine that a new food source appears, brightly colored red worms that are much more nutritious than the brown ones common in this environment. Selection will now favor both males and females who are preferentially attracted to the red worms. Now that females have the sensory/cognitive bias for red, when males evolve red epaulets, the females are already tuned in to seeing this color. Thus, while the females are searching for a mate, they spot these more conspicuous males with this mutation more quickly than the less conspicuous males, and even though this mutation might still exact a cost from predation, it is also generating benefits to the male because of increased matings. This is sensory exploitation, because when the males evolve the trait, there is already a hidden bias for that trait, in this case for the color red. This scenario has played out in the evolution of courtship displays in guppies.[29]

We have reviewed a handful of scenarios that can explain the evolution of female preferences: females get direct benefits from their mates that increase their number of offspring; females get indirect benefits from their mates by passing on "good" survivorship genes to their offspring; females get indirect benefits from their mates by producing sexy sons; and perceptual biases harbored by females are recruited into mate preferences when they are exploited by male traits. To make matters more complicated, these are not mutually exclusive scenarios, and they can all interact with one another.[30] Whereas Wallace and other critics of Darwin thought there were not plausible explanations for why females should show preferences for elaborate male traits, today there are a plethora of plausible explanations. The current controversy is over which of these hypotheses are more important in which species.

The Neural and Cognitive Bases of Mate Choice

Today, 150 years after Darwin's presentation of sexual selection theory and 50 years after Trivers's foundational paper that resurrected this theory, there is no question about the ubiquity of female mate choice in nature. In addition, the emphasis of much of the work in animal sexual selection has shifted from an understanding of the forces that can lead to the evolution of female mate choice to an understanding of the neural and cognitive bases of this behavior. Many of these studies have been quite successful in documenting the biological bases of aesthetic sexual preferences in animals.

Jacob von Uexküll introduced the concept of the *Umwelt*,[31] the main idea of which is that different types of animals reside in different sensory worlds. For example, unlike many birds, we cannot see in the ultraviolet; unlike bats, we cannot hear their ultrasonic echolocation calls; and unlike dogs, we have only very restrictive access to most of the odors around us. These sensory biases play an important role in the details of male sexual traits that females find attractive.[32] For example, the colors to which the eyes of surfperch are most sensitive have evolved to enhance prey detection in the kelp forests where these fish reside. These sensory biases were then targeted by the details of the courtship displays that males evolved to attract females.[33] Similarly, the túngara frog has evolved additional syllables, called chucks, to adorn its basic mating call, a whine, which stimulate an inner ear organ that had yet to be recruited for use, in communication among close relatives.[34] This added stimulation of the ear has two results: it leads to enhanced stimulation of the auditory centers in the brain, and female túngara frogs find whines with chucks more attractive than a simple whine.[35] Sensory biases are key components of the aesthetic preferences of females.

Biases in how we perceive the world are not restricted to sensory end organs—eyes, ears, and noses. There are also a number of cognitive biases that can influence an animal's taste for the beautiful. Many human psychophysical perceptions follow a power function known as Weber's law.[36] Our perception of the difference in the magnitude of two stimuli, for example, decreases as the absolute magnitude of the stimuli increases. We can more readily detect the small difference between a one-kilogram and a two-kilogram weight than we can detect that same difference between a twenty-five-kilogram and a twenty-six-kilogram weight. Weber's law also influences how animals weight the attractiveness between different stimuli. The túngara frogs mentioned above can add from one to seven chucks to their mating call. Females prefer whines followed by more chucks over whines followed by fewer chucks, but the strength of preference depends on the total number of chucks added to the call. The strength of preference between whines followed by two verses one chucks is much stronger than the preference between five versus four chucks.[37]

Weber's law also influences our perception of the attractiveness of human faces in at least a couple of populations of undergraduates in the United States. Gassen et al. morphed a series of faces from pairs of what were previously judged as an unattractive face and an attractive face.[38] The difference in perceived attraction between pairs of faces that differed morphometrically by 10 percent was much greater for two unattractive faces than it was for two attractive faces. These studies of frogs and humans indicate that the aesthetics of these two species does not vary linearly with the traits they are judging. These results have important implications for the tempo and mode of the evolution of beauty.

In animal studies, we often think aesthetic preferences are fixed for an individual's lifetime, while in humans, on the other hand, percepts of attractiveness of both men and women are often thought to be notoriously fickle. Pennebaker et al. showed that human perceptions of the attractiveness of the opposite sex vary as a function of "closing time."[39] Patrons at a bar were asked to rate the attractiveness of same- and opposite-sex patrons early in the evening and then again as closing time approached. Men and women both perceived opposite-sex patrons as more attractive later in the evening. When this study was repeated, and alcohol blood levels were measured and statistically controlled for, the same closing time effects were found.[40] One interpretation of these results is that to avoid cognitive dissonance, individuals change their perception of the attractiveness of individuals they would like to meet as the time available for social interactions decreases.

Animals face their own closing times. Female túngara frogs visit a chorus of males only on the night they are going to mate. If they do not choose a mate by the end of the night they still ovulate, expelling all of their eggs into the pond. It then takes the females another six weeks to develop a new set of eggs ready for fertilization. Lynch et al. tested females early in the evening with synthetic calls that had previously been shown to be unattractive to the females; these calls rarely elicited a response from the females—that is, movement to the speaker broadcasting the call.[41] The exact same calls, however, were very attractive to the same frogs later in the night, when the females were becoming desperate to find a mate. Both humans and female frogs are broadening their tastes for the beautiful as closing time approaches.

Peer pressure is another factor that can lead to unstable percepts of attraction. A simple experiment with a fish (a molly) showed that peer pressure is not restricted to humans. A female fish was placed inside a test tank with one male on each side of the tank. The males were separated from the focal female by a pane of glass, but the female could approach the males and be courted by them. When females were tested, they preferred courting with the larger of the two males. The female was then constrained to the middle of the aquarium in a glass cylinder, from which she observed the previously un-preferred male being courted by a female selected as the "model female." The model female was then removed, and the female-choice test was repeated; now the female switched her preference to the previously un-preferred male. The interpretation is that she was copying the mate choice of the model female.[42] Furthermore, the more attractive the model female, the greater her influence on mate-choice copying by the focal female.[43]

A number of studies have examined mate-choice copying or, more specifically, context-dependent attractiveness in humans. Typically, individuals rate a photograph of a person pictured with a member of the opposite sex as more attractive than a photograph of the same person alone. Recently, Gouda-Vossos et al. conducted a meta-analysis and showed there was strong evidence for this phenomenon in women but not in men.[44] Street et al. also demonstrated female mate-choice copying based on facial attractiveness, but they showed that the degree of copying was no different when analogous experiments were conducted in other domains, such as evaluation of art.[45] Their interpretation is that, at least in this instance, mate-choice copying is one expression of social facilitation and is not necessarily a psychological adaptation for mate choice. This does not lessen the influence of mate-choice copying in the evaluation of facial attractiveness and is yet another example of how cognitive biases in an individual's sexual aesthetics can have origins in domains outside of mate choice. Furthermore, these studies illustrate one role of social context in evaluating sexual beauty and suggest the importance of evaluating such phenomena across the breadth of cultural variation that is the hallmark of our species.

Summary

Darwin proposed his theory of sexual selection to explain how elaborate, sexually dimorphic traits used in courtship could evolve, despite being maladaptive for survival. Although sexual selection by male-male competition was readily accepted, the same was not true for sexual selection by female choice. The theory's strongest critic was Alfred Russel Wallace, the codiscoverer of the theory of natural selection. Initially, Wallace was skeptical of female choice in general, doubting that females would base consistent mate preferences on nuances of male courtship behavior and morphology. When he did grant some credence to female choice, he suggested choice was based on traits that indicated male vigor. Darwin, on the other hand, suggested that female animals have a "taste for the beautiful,"[46] that they possess sexual aesthetics not all that different from ours. His theory lay dormant for 100 years and was resurrected in the 1970s, primarily by Trivers's theory of parental investment. Since then, there have been hundreds of studies to demonstrate clearly the efficacy of female mate choice, thus validating Darwin's primary prediction about sexual selection. Disagreements still exist as to what causes the evolution of female mate choice, echoing the fundamental disagreement between Darwin and Wallace. There are now a number of studies supporting Wallace's utilitarian view that females acquire direct and indirect benefits from their mate choice. Other studies support Darwin's notion of sexual aesthetics by documenting the sensory, neural, and cognitive biases that underlie female mate preferences. The recent trend in the field has been to delve deeper into mechanisms of mate preference to further document the female's taste for the beautiful.

9

This View of Wife

Holly Dunsworth

Man is more courageous, pugnacious, and energetic than woman, and has a more inventive genius.

CHARLES DARWIN, *THE DESCENT OF MAN*, CHAPTER 19[1]

AT A GLANCE, Chapters 19 and 20 of *Descent of Man* look like Victorian Age–appropriate evolutionary explanations for skin and hair differences between the sexes and among peoples across the globe, and for how our naked, colorful skin developed out of our ape ancestry. But underneath Darwin's ambitious scientific contribution lies much more than surface beauty. This is Darwin begetting every caveman-inspired nugget of dating advice, every best-selling author's stance on innate gender roles, and every entertainer's sexist appeal to science. This is where Darwin first turned his concept of sexual selection loose on humans, launching the evolutionary narrative starring "the strongest and boldest men . . . in contests for wives" that dominates popular culture.[2]

In *Descent*, Darwin parlayed visible anatomical differences between sexes—like those in skin color, hair patterns, and height—into the evolutionary "logic" behind why Man and Wife perform differently, in matters of love, sex, parenting, cognitive feats, and seemingly everything else, according to the contemporary world views he continues to shape.

But pop culture has been slow to adopt new knowledge that has complicated and overturned old facts. Many of the novel insights on human evolution have come from women, despite persistent beliefs like Darwin's that, "with woman the powers of intuition, of rapid perception, and perhaps of imitation, are more strongly marked than in man; but some, at least, of these faculties are characteristic of the lower races, and therefore of a past and lower state of civilisation."[3] In *Descent*, Darwin even concludes, as if by scientific logic, that "man has ultimately become superior to woman."[4] For Darwin, women were wives, but men were so much more than husbands, an assumption seeding his science of sex differences. From his view of life, Darwin penned nature's seal of approval.

Here, in this chapter, we first update Darwin's science of skin, hair, and other anatomical variation. Many of Darwin's scientific offerings in Chapters 19 and 20 of *Descent* have crumbled under the weight of scientific progress. Likewise, some of his unscientific offerings have been exposed, and so, in the second part of this chapter, we confront his bias. And by "we," I mean you, the reader, and I, led by the women who speak throughout this chapter. Like each of us, Darwin was only himself, which is an insufficient lens for viewing the truth that we yearn to discover through science.

Since *Descent of Man*, 150 brief but productive years of humanity demand we find new ways to the truth about our evolutionary origins. Doing so is a fitting way to honor Darwin, who wrote about "false views" and taking "pleasure in proving their falseness; and when this is done, one path towards error is closed and the road to truth is often at the same time opened."[5] Thanks to hindsight, we know that Darwin held some false views, and, thanks to his blessing, we can take pleasure in proving their falseness.

The Science of Human Variation Has Evolved

Humans vary in pigmentation and pilosity. Ethiopians have much more pigmented skin than Estonians, and in every society, men are more pigmented than women. On the topic, Darwin wrote, "In certain races the women are said to differ slightly in tint from the men. . . . As the [African] women labour in the fields and are quite unclothed, it is not likely that they differ in colour from the men owing to less exposure to the weather. European women are perhaps the brighter coloured of the two sexes, as may be seen when both have been equally exposed."[6] Likewise, hair patterns vary by sex. No matter how denuded men are compared to other primates, women are even smoother and lack the conspicuous facial hair of adult males. About this ancient primate pattern, Darwin wrote,

> We have seen that certain male monkeys have a well-developed beard, which is quite deficient, or much less developed in the female. No instance is known of the beard, whiskers, or moustache being larger in the female than in the male monkey. Even in the colour of the beard there is a curious parallelism between man and the Quadrumana, for with man when the beard differs in colour from the hair of the head, as is commonly the case, it is, I believe, almost always of a lighter tint, being often reddish. I have repeatedly observed this fact in England; but two gentlemen have lately written to me, saying that they form an exception to the rule. . . . Women in all races are less hairy than the men; and in some few Quadrumana the under side of the body of the female is less hairy than that of the male. Lastly, male monkeys, like men, are bolder and fiercer than the females. They lead the troop, and when there is danger, come to the front. We thus see how close is the parallelism between the sexual differences of man and the Quadrumana.[7]

Those patterns both between and within populations, Darwin proposed, are due to sexual selection via local beauty standards for skin tone and hair growth. But while science still links the evolution of human skin color, fur loss, and hair patterns, as Darwin did, it is not how he imagined.

That people with deep ancestry in the tropical regions of the globe have some of the most pigmented skin is best explained by natural selection. This adaptation likely materialized after our hominin ancestors lost a significant amount of their protective fur covering. In its place, a

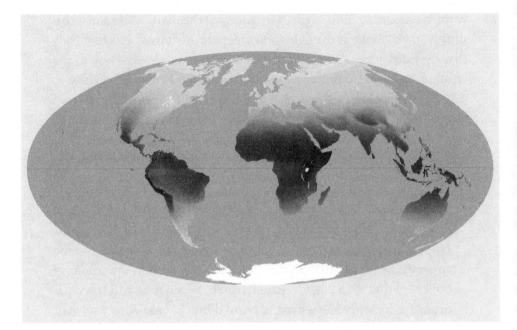

FIGURE 9.1. Skin coloration, caused by variation in pigmentation (melanin), grades with UV exposure in humans. (© George Chaplin)

melanin-rich epidermis shields the body from the sun's harmful ultraviolet (UV) rays.[8] As biological anthropologist Nina Jablonski has demonstrated, when maps of global annual UV intensity are laid over maps of the global pattern of human melanin variation, there is stunning alignment (Figure 9.1).[9]

For folks living in sunny, UV-intensive climes around the equator, there are numerous benefits to having a more pigmented epidermis. High levels of melanin in the skin can effectively prevent radiation from depleting folate, which is an important ingredient in many bodily functions, especially gamete production. Sperm with low folate have an increased risk of leading to debilitating and lethal outcomes in offspring. Epidermal protection against reduced fertility seems as if it would be especially sensitive to natural selection; it likely is and has been for a long time.

On the other end of both the melanin and UV spectra, at high latitudes, natural selection may also be the dominant explanation for human skin color. Our bodies require UV radiation to synthesize vitamin D, so some UV radiation is beneficial. In UV rich environs, skin with lots of pigment is able to welcome enough penetrating UV to carry out this process. In regions where there is less UV, and less chance of exposure to it, skin with less pigment allows UV to stimulate sufficient vitamin D production. With sufficient vitamin D, myriad bodily functions run smoothly, including the healthy development of the skeleton, which benefits successful reproduction. Dangerously low levels of vitamin D can lead to rickets (soft bones and skeletal malformation), among other complications.

Both skin-color extremes have strong adaptive explanations. In between the equator and the Arctic, that clinal or spectral continuous variation, from highly pigmented to depigmented skin, is maintained by gene flow connecting the populations and by local selection for intermediate UV levels in between the extremes. The evidence for this explanation for the evolution of skin-color variation in humans around the world far outweighs the evidence for Darwin's hypothesis of sexual selection via different beauty standards.

Researchers who are actively identifying and investigating the genes contributing to skin-color variation are quick to point out that it is not merely natural selection in specific UV environs that explains all regional variation in human skin color; it is also genetic drift.[10] This helps explain why some populations evolved to have similar skin tones via different mutations. Recently, a group led by geneticist Ellen Quillen summed up the state of the science of human skin color, its complex evolution, and its Eurocentric history:

> Even for genes that have been exhaustively characterized in European populations like MC1R, OCA2, and SLC24A5, research in previously understudied groups [i.e., non-European populations] is leading to a new appreciation of the degree to which genetic diversity, epistatic interactions [i.e., development in the environmental context], pleiotropy [multiple, often disparate, effects of one gene], admixture [gene flow between populations], global and local adaptation, and cultural practices operate in population-specific ways to shape the genetic architecture of skin color.[11]

For further context on human skin color, biological anthropologist Tina Lasisi writes, "Populations in Africa have more genetic diversity than non-African populations, yet this diversity remains underrepresented in studies of the evolutionary genetics of quantitative traits [like skin color]. . . . The inclusion of underrepresented populations is crucial for a more complete understanding of the evolutionary history of human variation."[12] The natural history of global skin-color variation is more complex than a single selection-based narrative, like that presented in *Descent*; experts now acknowledge and account for the cultural, political, and historical context that biased prior knowledge.*

Women generally have less melanin than men. To explain this patterned sex difference in human skin color, UV-thinking leads to a natural selection-based hypothesis too. As Nina Jablonski explains, it is "possibly because of the greater need of females to produce vitamin D in the skin to absorb and mobilize calcium during pregnancy and lactation." On whether Darwin's hypothesis still stands, she adds, "Sexual selection does not appear to have been a major influence on the evolution of human skin coloration, but it probably did increase the degree of sexual dimorphism in skin color in some populations."[13]

Additionally, differences in skin color between the sexes may come down to the hormones underlying reproductive physiology that also cause men's more vascularized upper dermis, greater amount of melanin, and thicker skin. Men experience a more intense tanning response, while women lose pigmentation more quickly.[14] Hormones are implicated here because of known changes to skin tone during puberty, across the life span, and during pregnancy, which occur because skin metabolizes hormones and responds to them. If further research demonstrates that sex differences in pigmentation are primarily by-products of hormonal interactions that differ between the sexes due to selection for their more direct but still adaptive role in male and female reproductive physiology, then neither sexual nor natural selection *for* skin color itself may be significantly responsible for sex differences.[15]

---

* And they are doing so without echoing Darwin's depictions of "savages" with their "monstrous" admiration of dark skin.

Darwin was right to emphasize that once sex differences exist, they can become targets of mate choice, preference, recognition, and attraction, and that those phenomena affect evolution by influencing what genes get paired up and pushed into the future. Still, if sex differences are now such targets, it does not necessarily follow that sexual selection ratcheted them up to where they are now.[16] Further investigation beyond a trait's present perception in a species is required in order for sexual selection to explain its origins.

One of the strongest testaments to Darwin's influence is that the mere existence of sex differences is enough for some scientists to presume that sexual selection brought them about. Take, for example, sex differences in human height. Sexual selection was Darwin's explanation:

> There can be little doubt that the greater size and strength of man, in comparison with woman, together with his broader shoulders, more developed muscles, rugged outline of body, his greater courage and pugnacity, are all due in chief part to inheritance from his half-human male ancestors. These characters would, however, have been preserved or even augmented during the long ages of man's savagery, by the success of the strongest and boldest men, both in the general struggle for life and in their contests for wives; a success which would have ensured their leaving a more numerous progeny than their less favoured brethren.[17]

Male body size remains tied to male strength, competitiveness, and aggression in discussions of the evolution of human sex differences. Darwin drew inspiration from the size, strength, and ferocity of silverback gorillas, and scientists in his wake have helped to build the case, which is now canon, that sexual selection explains sex differences in body size. Big males win greater mating opportunities by physically dominating competitors and mates, and by females preferring to mate with these big winners, so the thinking goes. This behavior is nearly always assumed to have caused the sex differences in size, rather than it being a consequence of the size difference or about something else entirely.

However, when the knowns and unknowns of skeletal growth are explored, it becomes difficult to imagine where sexual selection for tall men fits into the evolutionary story for sex differences in height. That is not to deny standards of beauty and attractiveness, and it is not to deny advantage that relatively greater height lends to physical and political achievement. However, sex differences in human height are at least as much about estrogen, ovaries, live birth, and lactation as they are about big strong men towering over the average Joes and winning over wives. When estrogen peaks during routine female development, the growth plates fuse, and teenage girls stop gaining stature. This estrogen peak at puberty is fundamental to ovarian development and crucial to the initiation of regular menstrual cycling. Typical teenage boys don't have ovaries or menstrual cycles, and so they typically don't have high levels of the hormones that cause bones to stop growing at the age that they do in girls. As boys continue to develop their long bones into their teen years, their growth plates age, and they eventually reach a stage when their relatively lower estrogen levels can affect growth-plate fusion too. Boys stop gaining stature, just a few years after girls, because there is nothing causing them to stop sooner, as there is in girls. Both male and female bodies depend on a delicate balance of estrogen (not too much, not too little) for gonad, genital, and gamete function. So, to assume that sexual selection could reduce estrogen in males in favor of an extended growth period may be to underestimate the estrogen needed to pass on those hypothetical estrogen-reducing genes to offspring. And, of course, bone growth is affected by numerous other factors, hormonal and otherwise, that are involved in multiple functions beyond body size. Underneath all the factors that explain human height variation, there are hundreds (and maybe thousands) of genomic connections. But, as of yet, there are no identified male-specific genes for male-specific biology of height; there is only common biology of skeletal growth shared with females, in which similar processes significantly controlled by estrogen play out differently in different bodies during development.[18] These issues, and the remaining gaps in our knowledge, don't apply just to human skeletal growth but to that of big competitive gorillas too.

Body size differences are not as simple to square with sexual selection as it seems. And yet, the sexual selection explanation prevails among professional and popular circles alike. Perhaps sexual selection is seductive because it matches a causal mechanism (male competition and female preference) with a familiar sociopolitical condition and its values. In the Darwinian context, Western notions of masculinity and patriarchy are justified because gendered behavior is deemed the driving force of evolution—a process that Darwin valued as progress and improvement, and our culture still does.

Differences in hormone levels that contribute to sex differences in skin tone and in skeletal growth may also explain differences in hair growth between the sexes. When estrogen-related processes are perturbed in women's skin, alopecia or hirsutism can result. In men, the reduction of testosterone coincides with cessation of facial hair growth and the onset of male pattern hair loss.[19] And, again, when such differences exist, they can be incorporated into gendered standards of beauty without necessarily making a significant impact on the evolutionary process. That is, there is no need for beauty standards to drive the sexes apart, because of the evolution of sexual reproduction and its effect on hormone levels.

Still, Darwin's hypothesis for *Homo sapiens'* fur loss was different. He posited that preferences for smoother females drove the evolution of fur loss in our entire species. While sexual selection has not been ruled out, it is challenging to empirically test, and it is also weak in comparison to present biological knowledge. There is a literature on tufts of hair in human armpits and pubic regions having to do with scent dispersal, given that we grow hairs where we also have sebaceous sweat glands (which ooze a milkier kind of odoriferous sweat than eccrine sweat glands).[20] But even if that function could be linked to evolution via sexual selection now, it does not tell us how the denuding and tufting began in the first place. The genomic divergence between human head and body lice is estimated to have occurred as far back as ~1 million years ago and provides a clue as to when hominin fur loss began, but it cannot tell us why or how.[21] Was it primarily due to sexual or natural selection, or to neutral processes and chance?

While the loss of a protective fur coat was indeed a loss, few support Darwin's view, that "no one supposes that the nakedness of the skin is any direct advantage to man; so that his body cannot have been divested of hair through natural selection."[22] On the contrary, natural selection is the winning supposition. Perhaps because of its fit with the UV-based skin-color hypothesis, the thermoregulation or body-cooling hypothesis is the most accepted explanation.[23] Less furry skin increases the effectiveness of cooling off through the evaporation of sweat, and this is adaptive for bipedal, big-brained (read: hot-brained) hominins who are active in tropical climes during the hot parts of the day. In this naked-in-broad-daylight niche, humans would also benefit from melanin-rich skin. Contrary to Darwin's assumptions, there are advantages to multicolored and naked skin,* which he was unable to imagine at his moment in history, through his eyes and from within his skin.

No One Can Shed Their Skin,
So Science Needs Everyone

Darwin literally expanded his horizons by sailing around the world. Then, from his home office, he broadened his mind by corresponding with other naturalists, scientists, and thinkers of his day. Criticism prodded him to create new experiences as well, like the time he bit the claws off a kitten to demonstrate that a clever baboon he described could indeed disarm its pet felid.[†] But no matter how much we grow our knowledge, none of us outgrows our bias, and that goes even for those giants like Darwin, on whose shoulders we stand.

Darwin was constantly weaving observations into explanations, using them to seek answers to questions and to form new questions and new answers. He wrote, "I am a firm believer, that without speculation

---

* Assuming that these traits *must* be advantageous is a whole other state of mind (a.k.a. adaptationism) for interrogating elsewhere.

† Darwin wrote, in the 2nd ed. of *Descent* (1874, p. 70), "A critic, without any grounds ('Quarterly Review,' July, 1871, p. 72), disputes the possibility of this act as described by Brehm, for the sake of discrediting my work. Therefore I tried and found that I could readily seize with my own teeth the sharp little claws of a kitten nearly five weeks old."

there is no good & original observation."[24] This reads as if he is excusing himself and his pen pal Alfred Russel Wallace of the potential errors that result from their efforts to understand nature. His intuition about his intuition was later backed by science.

Developmental psychologist Alison Gopnik says we're all born scientists and that, "nature has designed us to understand nature."[25] Thanks to many things, including our "explanatory drive," which is present even in babies, she explains, "we look beyond the surfaces of the world and try to infer its deeper patterns. We look for the underlying, hidden causes of events. . . . It's not just that we human beings *can* do this; we *need* to do it."[26] So, Darwin, Wallace, and everyone else are excused. We can't help being like this, and, anyway, there is no knowledge without explaining what we perceive. That is science, and it is possible only *with* humanity.

Darwin had a gifted explanatory drive, and he was aware of his humanity, his skin, but he harbored another defining trait of our species, blind spots:

> The chief distinction in the intellectual powers of the two sexes is shewn by man attaining to a higher eminence, in whatever he takes up, than woman can attain—whether requiring deep thought, reason, or imagination, or merely the use of the senses and hands. If two lists were made of the most eminent men and women in poetry, painting, sculpture, music—comprising composition and performance, history, science, and philosophy, with half-a-dozen names under each subject, the two lists would not bear comparison. We may also infer, from the law of the deviation of averages, so well illustrated by Mr. Galton, in his work on "Hereditary Genius," that if men are capable of a decided eminence over women in many subjects, the average standard of mental power in man must be above that of woman.[27]

Here is where Darwin's blind, explanatory drive sends his reason right off a cliff. Imagine someone arguing, *If two lists were made of the happiest children playing at a playground and the happiest children sitting in a church service, the two lists would not bear comparison*, and then daring to explain that difference with "Hereditary Happiness." Even though he was relatively progressive on race and gender issues, to Darwin white men were superior and, therefore, men were the most evolved humans, which in turn increased white male superiority because he valued "higher" states of evolution (which are fictions). Excusing one's explanatory drive is one thing, but it's something else entirely to excuse the systemic oppression of humans by narrating it away as evolved (read: good!) sex differences.

In science, as in all human endeavors, there are effective ways to forge ahead despite our limitations. These do not include willing our minds to be more objective. We cannot pretend as if we can peel off our person and our culture to reveal a wiser scientist underneath, preferably one of heavenly descent.* Yet that is what biologist Thomas Henry Huxley suggested we try, when he said, "let us imagine ourselves scientific Saturnians."[28] Decades later, in *Sociobiology*, E. O. Wilson wrote, "Let us now consider man in the free spirit of natural history, as though we were zoologists from another planet completing a catalog of social species on Earth."[29] Wilson's use of this trope as he introduced a new field of science is telling. Under the ruse of maximum objectivity and a hyper-Darwinian view of life, Wilson seems to absolve sociobiology of any of the harmful social and political consequences that result from its approach, which has seemingly infinite applicability.[30] Sociobiology's rise in the 1970s perpetuated old, and inspired new, captivating but biased (and worse) stories about human evolution and "human nature" that continue to grip popular culture.

Science is the human creation of knowledge. Prizing the completely objective scientific mind is foolish because such a mind does not exist. This myth is dangerous because it creates opportunity for scientists to

---

* Of anthropology—a field including both fellow travelers with, and ruthless critics of, sociobiology and its offspring evolutionary psychology—Jon Marks writes that it acknowledges, "the impossibility of studying humans as if we were not ourselves human and to reconceptualize the project as necessarily a biocultural one, infused with cultural values of greater or lesser transparency, but no less scientific for it. Or at least more scientific than pretending that you are a Martian." See Jon Marks, "Why Be against Darwin? Creationism, Racism, and the Roots of Anthropology," *Yearbook of Physical Anthropology* 55 (2012): 95–104.

exploit humanity. It is also detrimental to science because it prizes those perspectives judged to be objective over others, excluding so many from knowledge production, while elevating the "objective" perspective to fact.[31] As biological anthropologist Robin Nelson writes, "The idealization of an objective and apolitical science built on rational thought and deliberation has a face, and that face is white and male."[32] An uncritical belief in scientific objectivity does not just continue to dominate the profession, it influences the public perception of what counts as science and what is valued. Too often when these issues are raised, critics, who may even be scientists themselves, are labeled "anti-science" or "science deniers" or considered to be too politically motivated to be taken seriously as scientists. Just going in nontraditional scientific directions in a nontraditional body can result in a person's rejection from science. On these issues, indigenous-studies scholar Kim Tallbear writes, "Being differently situated is advantageous for producing different insights but has its risks. When one fails to exemplify a white Western often straight and masculinist gaze that is ironically seen to embody 'objectivity,' or if one researches too close to home, one gets accused of bias."[33] The persistent myth of an unblemished science is what tricks us into believing that Darwin's ideas, like his harmful beliefs about women, can be challenged only with science and, ironically, by those scientists deemed to be objective about such issues: men.

For both natural and sexual selection, Darwin's view of life was couched in competition, especially as he saw humans. But if he were someone else, he could just as easily have been biased toward cooperation. Perhaps this peek at his personality in a note to fellow naturalist Alfred Russel Wallace helps explain his bias:

> I infinitely admire & honour your zeal & courage in the good cause of Natural Science; & you have my very sincere & cordial good wishes for success of all kinds; & may all your theories succeed, except that on oceanic islands, on which subject I will do battle to the death.[34]

"Darwinian" has become a synonym for competition. While it may be fitting, it is misleading about evolutionary theory both in Darwin's time and now. Not only is cooperation fundamental to evolution,[35] some evolutionary scientists believe it to be even more important than competition.[36] Competition has dominated the scientific narrative; yet, whether it earned that dominance via scientific evidence is not entirely clear.

True to a "Darwinian" view of life, scientists often act and write as if evolutionary narratives compete for the truth. But as we have seen with the evolution of skin-color variation and sex differences in body size, that is not how the evolutionary truth works in the context of infinite connectedness and constant change. In biology, complexity is the norm, yet in science, simplicity is valued, as convention. However, nature is not as prone to domination by a single process as our narratives are.

Complicating or potentially overturning Darwinian views of competition and dominance can take thousands of hours of careful primatological observations. Rebecca Lewis has demonstrated just how complex power dynamics can be in primates, especially with regard to females. Her work has also revealed just how difficult competition and dominance are to study, because of issues of quantification and human bias. From her fresh perspective, Lewis has observed how dominance can be expressed differently between the sexes. She noticed that "refusing to act is a means of power that is often overlooked in primate studies and may be a more common means of intersexual power used by females than is aggression."[37] This is the case in primate species with considerable sex differences in body size, in which males are often assumed to have disproportionate power by observers who have ignored all but aggressive displays.

In the context of her own work on female-female competition, primatologist Sarah Hrdy illuminates why new perspectives like Lewis's are so crucial to progress in evolutionary science: "There has been a prevailing bias among evolutionary theorists in favor of stressing sexual competition among males for access to females at the expense of careful scrutiny of what females in their own right were doing."[38] Hrdy is writing about nonhuman primate behavior here, but she is acknowledging, too, the bias that built and favored "Man the Hunter"—the paradigm that researcher Sally Linton famously exposed and shifted in her paper

"Woman the Gatherer."[39] Work like Linton's, Lewis's, and Hrdy's has accumulated evidence for, and inspired more focus on, the crucial evolutionary roles of female primates. With such an expanded view, suddenly the widespread patriarchal system in our own species is harder to explain as some biologically destined circumstance, propelled by evolutionary momentum, that we're powerless to destroy.

Humans exhibit more extensive male dominance and male control of female sexuality than is shown by most other primates.[40] But interpreting this as selection for dominance is problematic, considering that, in humans, males are more similar in body size to females than is the case in many other primate species. Human males experience relatively lower male-male competition for mates as well. Of natural and sexual-selection-based explanations for male dominance and related behaviors, Lewis writes, "They start with the implicit assumption that the sexes should be different, implying that both nonaggressive males and aggressive, vigorous, and/or powerful females are *deviations* from the expected primate or mammalian conditions."[41]

Sex differences in body size and other traits, such as canine tooth size, do play a role in sex differences in dominance and power, but scientists are increasingly skeptical that they are caused by selection for dominance. As Lewis writes, "Relationships cannot evolve. Individual qualities that might influence intersexual power (e.g., body size, canine size, . . .) can evolve, however. Comparisons of closely related [species] can provide insight into which individual traits might have *led to* changes in intersexual power."[42]

Broadening evolutionary approaches to include both sexes, to keep up with cutting-edge knowledge in other fields to synthesize with one's own, and to include closely related species can be enlightening. So too can escaping the mindset that humans are just another ape. For example, male chimpanzees can be aggressive with prospective female mates. Of these sexually threatening apes, journalist Angela Saini writes, "If a scientist had only ever studied chimpanzees, he or she might conclude that this is the natural order of life for the great apes [including humans]. It's alluring and easy to draw parallels between patriarchal humans and macho male chimpanzees."[43]

Indeed, this is how so much pop culture evolutionary thinking plays out. But as anyone versed in ape sexuality knows, bonobos behave differently, despite being just as closely related to us as chimpanzees are (these two distinct apes are sister species). Darwin did not live to learn of science's discovery of the existence of bonobos in the early twentieth century or of the discovery of their close relation to humans with the advancement of molecular methods. If we choose to narrate human evolution according to the bonobo storyline instead, then we depart from Darwin's. Female bonobos attack males, have frequent sex, do not seem to be very choosy, coy, or reserved, and appear instead to be just as promiscuous as males. And in both bonobos and chimpanzees, depending on the population, females hunt and forage for animal prey as much or more than males do. So, we could look to some primates, as we know them now, and confirm Darwin's assumptions that "male monkeys, like men, are bolder and fiercer than the females. They lead the troop and when there is danger, come to the front."[44] But we could also counter that with many other primate species that are not characterized by male dominance and male-only aggression or power.

As one of the earliest scientists to overturn Victorian assumptions about female primates, Sarah Hrdy warned us all: "Chauvinists of both sexes have dipped into the primate literature to document their positions."[45] That we can string the same data together with completely different narratives is demonstration not of nature's duplicity but of its complexity, as is increasingly understood by scientific progress. But this progress was delayed.

In the late nineteenth and early twentieth centuries, American opposition to female higher education prevented women from entering the same knowledge-creation business that built, claimed, and perpetuated Darwin's work—which conveniently suggested that women weren't suited for higher education. About the powerful effect scientific and lay interpretations of *Descent* had on women at this time, historian Kimberly Hamlin writes,

> To evolutionists . . . women's inferior intellect was not a flaw but rather an evolutionary necessity for the creation of healthy

offspring . . . since maternity took up [so much energy] intellectual equality was not a goal to be pursued because it would lead to the extinction of the species. . . . To those men steeped in evolutionary discourse and the attendant pride in being at the pinnacle of all living things, women going to college threatened to minimize sex differentiation [which Darwin asserted to be an achievement of the most advanced species], thwart evolutionary advancement, and diminish white racial superiority.[46]

Just after its publication, readers interpreted *Descent of Man* as a call to action to contribute to the evolutionary process by aiding selection's improvement of our species. Where there was once God's plan that we must carry out, now there was selection's. Talk about stringing the same data together with two different narratives. From this view, cutting-edge science justified limiting the freedom of all but upper-class white men. Women evolved to be wives (and not scientists or scholars) and to carry out evolution's plan. Natural and sexual selection conveniently favored what society already did. The scientific value of *Descent* is impossible to untangle from the oppression that it inspired.

Perhaps more than any other science, evolutionary science is a "collection of stories about facts."* These stories are difficult to separate from the facts and, indeed, become *the* facts without as much burden of proof placed on them as some facts. For example, men are on average larger than women (fact) because of large, winning male ancestors (fiction—i.e., hypothesis or possible explanation—which is far more difficult to establish as fact). Anthropologist Venla Oikkonen describes the power of narrative as "one of the primary mechanisms through which assumptions about biological gender differences are produced as scientific fact in popular discourses. If narrative naturalizes ideas and experiences, then narratives of gender turn gendered assumptions into

---

* According to Adam Gopnik, David Dobbs, in his book *Reef Madness: Charles Darwin, Alexander Agassiz, and the Meaning of Coral* (New York: Pantheon, 2005), described Darwin's subsidence theory for coral reefs being old volcanoes as "Herschelian." John Herschel, Victorian astronomer and philosopher, "pioneered the notion that science is a collection of stories about facts, not a mere collection of data dumps," writes Gopnik. See Adam Gopnik, *Angels and Ages* (New York: Vintage, 2009), 86.

gendered facts."[47] Gender is culture, and culture has a talent for writing itself out of the story, for naturalizing itself, in deference to its preference for nature.

Metaphors are fundamental to good stories, and natural and sexual selection are powerful examples of both.[48] Along with metaphors, our stories seem to require characters with agency, for which selection has served as metaphor.[49] Where evolution by natural selection is not the "agent" in a tale, then sexual selection makes an appealing story, because it is actually about attractive and aroused male and female agents competing for and choosing sex partners! Sex stories sell. Novelty does too, and so our sexual selection narratives have evolved since Darwin's day. In an insightful critique of Wilson's *Sociobiology*, which was published in 1975, anthropologist Anne Fausto-Sterling writes,

> In a century's time the image of the female consumer shopping for the best mate has replaced [Darwin's vision] of the protectress of beauty and refinement. Missing, too, is the image of two sexes cooperating to bring forth the next generation. Instead Wilson presents courtship as a contest in which males and females are fundamentally at odds; his interest is to reproduce regardless of how fit he is, even if success entails trickery, and hers is to increase her fitness by exposing deceit, choosing for herself the best possible mate.[50]

For those who understand the difference between facts and the stories we build around them, for those who see the bias and prejudice that is infused in so much evolutionary thinking, it is painfully clear that as individuals we are not capable of telling all the stories in science all by ourselves. Try to imagine Darwin summing up an afternoon spent among the primates at the San Diego Zoo, as anthropologist and writer Wednesday Martin recently did, upon finding her way to a hidden exhibit: "It occurred to me that the bonobos might be tucked away because, with all their screwing and sucking and scissoring and shit eating, they are not exactly rated-G great apes. Our very closest relatives are far from family friendly."[51]

Descent of Man was far from family-friendly, even by 2021 standards, and so is Martin's book. Yet, it is largely thanks to candid and diligent scientific study of animals like bonobos that we can bust Darwinian myths that women (but not men) are naturally passive, coy, and sexless, while men (but not women) are naturally big, strong, arrogant, sex machines. In one of the most recent endeavors to expose just how tenuous and otherwise flat out wrong many of science's and society's assumptions about human sexuality have been, Martin jokes about women's libidos to make a serious point: "What seems beyond dispute is that women are . . . super freaks. . . . We are sexual anarchists . . . we might quite fairly be described as the largest group of perverts in America."[52] This new evolutionary script inspires a question: If people of genders, sexualities, races, ethnicities, classes, cultures, and life experiences different from Darwin's, if women like those highlighted in this chapter, had been included in the scientific pursuit of knowledge since the beginning, would there have been a *Descent of Man*?

An Evolution with and for Everyone

Fossil hunters know about light—how it can illuminate teeth and bones in the morning but hide them in plain sight after lunch. As told in *Descent of Man*, Darwin's "light . . . on man and his history"[53] was only a narrow band within all visible wavelengths. If we were seeing human evolution with light's full spectrum today, then Chapters 19 and 20 of *Descent of Man*, where Darwin proceeded in relative darkness, would be more artifact than domineering fact.

Part of the trouble is that "light" tends to privilege the hard sciences and their male-dominated traditions at the expense of other important knowledge. So, science continues to be deemed necessary to counter Darwin's science-less views on human differences and their sociocultural implications—for example, to demonstrate that women and people of color are not inferior, less evolved, closer to nature, less deserving of opportunity, influence, and admiration than white men. While many scientists and scholars have met Darwin's bias, racism, and sexism with science, science is not the only, or even a necessary, way to demonstrate humanity. That is partially because science does not yet represent humanity.

The culture of science decides whose views and voices are worthy of expanding or overturning scientific views and whose are not. Ironically, this tradition inhibits scientific progress. Just being alive in the future relative to when the ideas we have inherited were conceived is not enough for progress. Scientists must actively reflect on their limitations and confront their biases if they truly wish to understand the world. And science, collectively, must comprise the diverse humanity of this world we wish to understand.

With a deeply reflective humility, we can appreciate those shoulders we stand on, while remaining painfully aware that they belong to humans, not giants. Humans are metaphorically and literally bound in our skin. The outer observable layer we recognize as ourselves is actually the dead remains of deep, active epithelial cells. Our identities and our resulting outlooks are more about our past than we often realize—this goes for our personal biographies as well as the evolutionary histories we narrate.

Everything, our species and our science, is because of what came before. This is why science is a sort of martyrdom; integral to our collective progress is the revelation of our personal stupidity and prejudice, our tendency to stick to old and ingrained thinking, unquestioned. It's largely thanks to Darwin that evolutionary science got so far beyond him to the present. The most fundamental aspect of selection is also true for science; they both tolerate the change they require. While there has been 150 years' worth of scientific light on human evolution, and that has meant that the story has changed all along, it is still not a narrative that everyone can embrace if they want to. But universality should be the priority of the writers of our species' shared origins story. The unscientific bias in *Descent of Man* that Darwin legitimized as science, and that plagues us all, may be unavoidable, but it is surmountable. The stories we tell about the facts of evolution are in dire need of diversity. The way to get there is not merely to be more correct or less biased than Darwin, and it is not merely to be the best version of ourselves; it is to be proactively bigger than any one person can possibly be.

Thinking hard all by yourself from your own skin is a good start, but it's only as good as the expanse of your lived experience and the range of your enlightenment. Just as much as everyone else, scientists need to broaden their spectrum, and they must find ways to enable all those lights to shine directly from their sources, unfiltered. For far too long, science has excluded the lives that could make this view of life even dare to approximate the truth. What's more, we owe it to our species to break the link between Darwin's inchoate offerings and their perceived social implications. Outdated science-inspired narratives alienate people from our shared-origins story, making it difficult for many people to claim it for themselves. Inclusivity, all of humanity, is the only path toward a less false view of life, with everyone and for everyone.

That's just one wife's view. What's yours?

10

Dinner with Darwin: Sharing the Evidence Bearing on the Origin of Humans

Ann Gibbons

I have given the evidence to the best of my ability; and we must acknowledge, as it seems to me, that man with all his noble qualities, with sympathy which feels for the most debased, with benevolence which extends not only to other men but to the humblest living creature, with his godlike intellect which has penetrated into the movements and constitution of the solar system—with all these exalted powers—Man still bears in his bodily frame the indelible stamp of his lowly origin.

CHARLES DARWIN, *THE DESCENT OF MAN*, CHAPTER 21[1]

ON A DRIZZLY Saturday morning in September 2015, a few dozen anthropologists, archaeologists, and I boarded a bus at the British Museum in London and headed an hour or so into the Kent countryside. We were making a pilgrimage, of sorts, to two shrines of human evolution—the gravel pit where the earliest known hominins in England were found, in

1935, in the village of Swanscombe, and then to Down House, in the rural village of Downe, where Charles Darwin lived for forty years. After several days of talks and many hours of PowerPoint presentations at the annual meeting of the European Society for Human Evolution, many of us were eager to get out of the museum's basement conference rooms and into the countryside, even if it meant mucking around muddy fossil beds in the rain. Over lunch at the George & Dragon pub in Swanscombe—and pints of aptly named "Caveman" and "Neolithic" ale—some of us also wondered whether, by visiting Darwin's habitat, we would find clues about how he formulated his ideas on human evolution.

When we arrived at Down House, the rain had stopped, and we fanned out over his eighteen acres of grounds, restored to appear much as they would have when Darwin lived here, from 1842 until his death, with his wife, Emma Wedgwood, and their seven children (ten births) and assorted pets and livestock.[2] As we explored the lush back lawn, we came upon the famous "wormstone experiment,"[3] in which Darwin and his children showed that given enough time, earthworms can remodel landscapes in significant ways. As we toured his orchard, we learned about his "weed experiment," in which he monitored the survival rate of seedlings when confronted with predators. After clearing a plot of land in January of 1857, Darwin then counted the number of seedlings that sprang up in the spring and summer (marking each with a small length of wire). Later, at regular intervals, he checked how many had survived—by August, he noted, from a total of 357 germinated seedlings, 205 had died, "chiefly by slugs and insects."[4]

As we toured his greenhouses, we saw carnivorous Venus flytrap plants (*Dionaea muscipula*), like the one that Darwin tried to trigger to snap shut with a "piece of very delicate human hair, 2½ inches in length,"[5] from his wife Emma's head. When he waved it over the plant's spikes, or filaments, it didn't close. But when he cut it into a more rigid inch-long piece and probed the filaments slowly, it did snap shut. Darwin concluded that, if "a minute insect alights with its delicate feet on the glands of *Drosera*, it is caught . . . and the slight, though prolonged pressure, gives notice of the presence of prey."[6] Those observations proved crucial to the formulation of his ideas about how different species of flowers have evolved to attract specific insect pollinators. They also amused Emma, who wrote: "At present he is treating *Drosera* just like a living creature and I suppose he hopes to end in proving it to be an animal."[7]

What quickly became evident during our tour at Down House was that the gardens and greenhouses were Darwin's laboratory for exploring how natural selection worked in plants, animals, and insects. But what about in humans? In Darwin's study, we could see the horsehair armchair where he wrote *The Descent of Man*, and we could imagine him stooped over his cloth writing board, speculating that humans were more closely related to African apes than Asian ones. Beside the chair, on a drum table, were casts of limb bones of animals, specimen bottles, and a microscope that Darwin used to peer at plant and animal specimens.

But there were no casts of hominins or primate bones or charts of human anatomy. Indeed, the absence of fossils of humans, apes, or monkeys at Down House was a reminder of just how little actual hard fossil evidence Darwin had to work with when he wrote *Descent of Man*.

Darwin relied instead on a lifetime of observations,[8] from his travels on the HMS *Beagle* to his own research on sexual selection and on the similarity of human and primate facial expression—which he observed by climbing into the cage of an orangutan named Jenny when he was a young man of twenty-nine so he could see her up close.[9] He also deeply depended on his correspondence and conversations with other researchers. Portraits over the fireplace in his study reminded us of his particular regard for the famous geologist Charles Lyell and the botanist Joseph Dalton Hooker, director of the Royal Botanic Gardens at Kew, and their profound influence on his work.

It also was Lyell, along with the paleontologist Hugh Falconer, who showed Darwin the only fossil of a hominin he would ever see—that of a Neanderthal skull discovered in 1848 in Gibraltar, as noted in the preface of this book. Darwin had known about the Gibraltar 1 skull, as it is known, ever since its discovery, but the significance of the find was not understood—it was described simply as "a human skull from Forbes'

FIGURE 10.1. Darwin's dining room at Down House, in Kent, United Kingdom.
(Photograph by Jeremy DeSilva)

Quarry" by the Gibraltar Scientific Society in 1848.[10] After miners working in Germany's Neander valley discovered another partial skull and
other bones of Neanderthals in 1856, scientists wanted another look at
the Gibraltar skull, and so it was brought to England in 1864 to a meeting of the British Association for the Advancement of Science and eventually into Darwin's hands.[11] Although the biologist Thomas Henry
Huxley, Darwin's close friend, thought it was an ancient form of human,
it's not clear what Darwin himself thought about it.[12]

The skull must have been the subject of discussion at Down House,
however, where Lyell and Hooker were frequent guests. Indeed, as we
walked through the dining room, it was tempting to imagine the conversation at the dinners those friends shared. What if Darwin could
have traveled in time to that day in 2015, to also host the remarkable
group of researchers passing through his dining room, where the table
was still set (Figure 10.1) with the same Wedgwood "Water Lily"
botanical-pattern service in earth tones that had been given to Charles

and Emma by Darwin's mother, Susannah Wedgwood? What would Darwin have wanted to know if he could have spoken with these anthropologists and archaeologists 133 years after his death? If they had that chance, what would they have told Darwin about what we've learned about human evolution since he and his guests had dined on Emma Darwin's beef collops (slow-cooked steaks of beef rump in a sauce of onions, soy sauce, and pickled walnuts) and burnt rice pudding.[13]

Chris Stringer, a paleoanthropologist at the British Natural History Museum who helped organize the expedition to Down House, would tell Darwin that he was right when he predicted in *Descent of Man* that our closest living allies are African apes rather than Asian apes. Darwin had said it was "useless to speculate on this subject,"[14] but that didn't stop him or his colleagues from conjecture. Huxley had enthusiastically promoted the view that humans should be put in the same family as African apes, while the prominent German biologist Ernst Haeckel thought that Asian apes (orangutans and gibbons) were closer relatives of humans.[15] "I guess Darwin would love to hear about our relationship with the great apes, and which was closest to us—chimpanzees, and of course two species [the bonobo wasn't known to science during Darwin's time]—and the nature of our common ancestor (not that we can answer that one clearly, even today)," says Stringer. "We forget how cautious he was in 1871."[16]

Adrienne Zihlman, an anthropologist and emeritus professor at the University of California, Santa Cruz, would be the expert at the table on the skeletal anatomy of apes and so would have shared with Darwin evidence from anatomical, behavioral, and genetic studies of apes—and explained how it has shown not only that chimpanzees are our closest relatives but also that fossils of the earliest hominins have shown that the common ancestor of African apes and humans must have had a small body, just as Darwin speculated—more like a chimpanzee's than a gorilla's. Although he admitted that "we do not know whether man is descended from some small species, like the chimpanzee, or from one as powerful as the gorilla," he thought it most likely that "an animal possessing great size, strength, and ferocity, and which, like the gorilla, could defend itself from all enemies, would not perhaps have become social; that this would effectually have checked the acquirement of the higher mental qualities, such as sympathy and the love of his fellows. Hence it might have been an immense advantage to man to have sprung from some comparatively weak creature."[17]

This common ancestor of African apes and humans was not exactly like a chimpanzee, because chimpanzees also have been evolving for millions of years since our ancestors' paths diverged, and they have acquired many derived traits, such as knuckle-walking, in response to the habitats where they lived. Darwin predicted a different evolutionary scenario than the all too familiar (and erroneous) trope of a chimpanzee slowly evolving into a human when he wrote, "But we must not fall into the error of supposing that the early progenitor of the whole Simian stock, including man, was identical with or even closely resembled any existing ape or monkey."[18]

But fossil evidence and behavioral studies suggest that the earliest hominins must have lived much like chimpanzees in the tropical forests in the heart of Africa, perhaps in the Congo Basin, where all three species of African apes are found today—gorillas, common chimpanzees, and bonobos.[19] Zihlman would tell him that both fossil and genetic studies also suggest that the ancestors of the earliest hominins branched off from the ancestors of chimps perhaps 7–8 million years ago,[20] moving east to the Great Rift Valley of Africa, where they came down from the trees to forage for food on the forest floor and to occasionally hunt monkeys or patrol their territory—and that eventually, by 6 million years ago, the upright-walking ground apes ventured out into more open wooded grasslands to find food or mates. Precisely why this new form of locomotion was advantageous to our ancestors remains unknown, although that hasn't stopped researchers from speculating.

Meanwhile, the parent population in the Congo eventually separated into two different groups of chimpanzees, about 1.5–2.6 million years ago, with common chimps (*Pan troglodytes*) heading west, where they competed with gorillas, and bonobos, or pygmy chimps (*Pan paniscus*), heading east and south, where they adapted to a variety of habitats.

Zihlman would also argue that "the evidence is stronger than ever" that bonobos, which are habitually less aggressive than common chimps, are a better prototype for the common ancestor than gorillas or the common chimpanzee. While debate continues about which African ape is the best behavioral model (or even *if* chimpanzees are good models for the common ancestor), Zihlman says that pygmy chimps' body size, sexual dimorphism, and other traits are closer than gorillas' to those of *Australopithecus* (from Latin *australis*, meaning "southern," and Greek *pithekos*, meaning "ape").[21] As detailed in Chapter 4 of this book, this is a genus of hominins whose best-known member is Lucy, the petite partial skeleton of an *Australopithecus afarensis* who lived at what is now Hadar, Ethiopia, about 3.2 million years ago—and whose species is a leading candidate for direct ancestor of early members of the genus *Homo* and, eventually, humans.

On the subject of *Australopithecus*, paleoanthropologist Leslie Aiello, president emerita of the Wenner-Gren Foundation in New York, would love to see Darwin's reaction to the hard evidence for human ancestors from the fossil record today. Given that he knew of only one putative human ancestor—the Gibraltar Neanderthal—she thinks he would be "astonished" to learn that the human family tree now includes more than twenty species of hominins and that the oldest fossils of putative hominins date back 6–7 million years.[22] "In his time, there was really no fossil evidence," says Aiello, who had taken many colleagues to Down House over the years she was on the faculty of the University College London. "He would be astounded at the variety and time depth." But how those many species of hominin—and those yet to be discovered— are related to one another, and which ones are on the direct lineage to *Homo sapiens*, remains a mystery.

Francis Thackeray, a paleoanthropologist at the University of the Witwatersrand in Johannesburg and former director of the Transvaal (now Ditsong) Museum in Pretoria, would particularly delight in telling Darwin how the first African fossil of a hominin was found, in 1924, by Australian anatomist Raymond Dart, who discovered "a little fossil called the Taung child," an apelike creature that lived 2.5 million years ago in what is now South Africa.[23] Dart would name it *Australopithecus africanus,* but it would take him more than a quarter century to convince most scholars that it belonged in the human family. Dart's claim was rejected by the scientific community in part because fossils found earlier in Indonesia (the famous "Java Man," which would later be recognized as *Homo erectus*) seemed to be better candidates—and because preconceptions about the earliest hominins required that a "missing link" walk upright and have a big brain. But over the years, as fossils from at least three different species of *Australopithecus* eventually were found in caves in South Africa in the 1930s and 1940s, including a remarkably complete skull of a female known as Mrs. Ples (whose species lived 2.1–2.5 million years ago near the Sterkfontein Caves), more and more scholars began to recognize that these small-brained creatures that walked upright were indeed members of the human family rather than ancestors of African apes. "Your words were prophetic," Francis Thackeray would say to Darwin today. "Does that cheer you?" (Thackeray's cousin, William Makepeace Thackeray, employed the same nurse, Jessie Brodie, who took care of Darwin as he wrote *Origin of Species* and reportedly nudged her charge, "Hurry up now, Mr. Darwin, you must do what Mr. Thackeray does and get down to writing in earnest to finish the book.")[24] If Francis Thackeray were to speak with Darwin today, he'd raise a glass: "Let's propose a toast to *Australopithecus,* and to fossils nicknamed Lucy, Mrs. Ples, and the Taung Child."

It wasn't until the 1950s, however, that most scholars finally recognized Africa as the birthplace of the human family. That happened after Mary Leakey took a walk with two of her dalmatians at Olduvai Gorge in Tanzania and spotted a bit of jawbone poking out of the sediments on July 17, 1959, 100 years after the publication of Darwin's *Origin of Species.* Noting its robust jaw and teeth, she and her husband, Louis Leakey, gave the fossil the nickname "Nutcracker Man" and the formal name *Zinjanthropus boisei* (also known as "Zinj").[25] Although Louis Leakey called Zinj "the connecting link between South African near men (like the Taung Child) and true man as we know him,"[26] the Zinj skull is recognized today as *Australopithecus boisei* and seen as more of a cousin whose lineage went extinct than as a direct ancestor of humans. Indeed, researchers have since learned that there were many branches on the

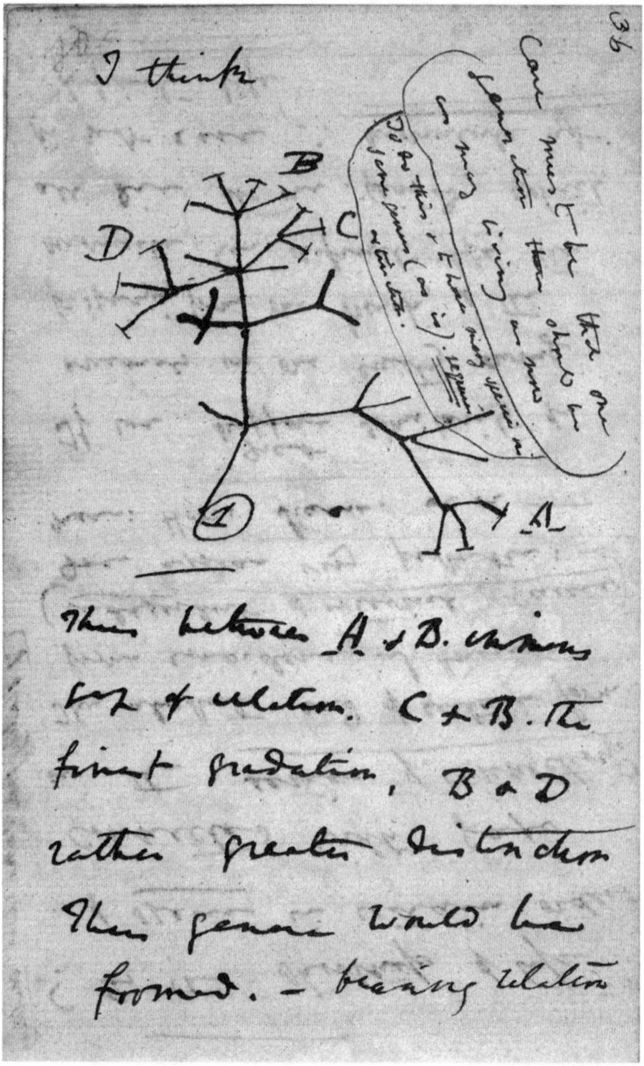

FIGURE 10.2. Darwin's famous "I think" tree, drawn in his 1837 Notebook B. It reads:
"I think. Case must be that one generation should be as many living as now.
To do this & to have many species in same genus (as is) *requires* extinction.
Thus between A & B immense gap of relation. C & B the finest gradation,
B & D rather greater distinction. Thus genera would be formed.—bearing relation"
("to ancient types with several extinct forms" continues on next page).
(Courtesy of the Cambridge University Library)

human family tree, with several types of hominins alive at any given time until relatively recently, rather than a single parade of one type of ancestor evolving into another in succession. This tree with many branches fits with Darwin's famous sketch of a tree of life in his notebook in 1837, in which he shows how a genus of related species might originate by divergence from a single starting point (Figure 10.2).

Within a few years of the discovery of Zinj, for example, the Leakeys' team found a hominin that lived at roughly the same time as Zinj but, with a slightly larger brain and smaller cheek teeth, better fit the bill as a human ancestor. This second skull and humanlike hand bones were found between 1960 and 1963 near stone tools, which suggested to Louis Leakey and the famed paleoanthropologist Phillip Tobias and anatomist John Napier that it could manipulate objects with precision— hence they put the species in our genus *Homo* and named it *Homo habilis*, or "handy man."

Just as important as finding out *where* the earliest hominins arose, however, was finding out *when* and *how* they arose and how they adapted to changes in their habitats. Aiello explains how this became possible with the development of new tools and new disciplines to help put the bones, stone tools, and artifacts into perspective. She thinks Darwin would be "astounded at what we can get out of the fossils now—ranging from the ancient DNA to the isotope analysis to the dating," she says. "Spinning off his oft cited aside . . . that 'In the distant future I see open fields for far more important researches. . . . Light will be thrown on the origin of man and his history,'[27] he would be overwhelmed at how bright that light would become!"

A revolution in dating of fossil sites in the 1950s and 1960s, for example, finally gave researchers a precise way to calculate the age of the fossils found amid volcanic sediments and to estimate roughly when key adaptations took place in human evolution, from upright walking to tool use. Louis Leakey had thought that Zinj lived 600,000 years ago at Olduvai, and geologists at the time thought that the Pleistocene—the age of humans—began about 1 million years ago instead of 2.55 million years, the time frame used today.[28] Employing a new radiometric dating method developed by physicists at the University of California in Berkeley in the 1950s to measure the radioactive decay of different isotopes of potassium and argon in volcanic sediments above or below a fossil—or both—geologists dated the fossil beds at Olduvai and found that the Zinj skull and *Homo habilis* were both about 1.75 million years old (currently dated to about 1.8 million years).[29] The new date made them three times older than previously believed.

The Leakeys' success at Olduvai inspired a new generation of researchers from many disciplines—anthropology, archaeology, geology—to travel to Africa to search in earnest for new fossils. New discoveries in the 1960s and 1970s helped researchers sketch out a rough time line for when hominins first appeared. By the mid-1970s, the discovery of Lucy and other members of her species at Hadar in Ethiopia had pushed back the origins of the earliest members of the human family to at least 3.2 million years.

By the year 2000, that date had receded even further into the past, with the discovery of new hominin species that lived between 6–7 million and 4 million years ago. These purported hominins included *Sahelanthropus tchadensis* from the Djurab Desert of Chad, *Orrorin tugenensis* from the Baringo Basin of Kenya, *Ardipithecus kadabba* and *Ardipithecus ramidus* (called "Ardi") from the Middle Awash of Ethiopia, and *Australopithecus anamensis* from Kanapoi, west of Lake Turkana, in Kenya. With their small bodies and brains and strange upright gaits, these hominins didn't look like us. As paleoanthropologist Tim White of the University of California (whose team found *Ardipithecus*) has quipped: "You wouldn't invite them to dinner."[30]

These fossils have shown that the earliest members of the human family still spent a lot of time in the trees, sleeping in nests safe from predators at night and picking seeds and berries by day. But when they came down to the ground, they walked upright and didn't put their weight on their knuckles, as African apes do. When they bared their teeth, the males revealed canines smaller than those in chimps. But this was not because they were already using stone tools and didn't need such sharp canines to tear meat and chew it, as Darwin suggested.[31]

Instead, the Taung Child, Mrs. Ples, Lucy, and the rest of the australopithecines, with their small brains and smaller canines in males, showed that upright walking and smaller canines came long before a big brain and tool use. Today, we know that bipedalism is one of the first defining traits of membership in the human family rather than the lineage of an African ape.[32] If Stringer and others were to lay out on Darwin's table a parade of new fossils of the feet, pelvises, and lower limbs of different hominins, ranging from Ardi to Lucy to "Little Foot" and "Karabo" (members of *Australopithecus* from South Africa), they could show him the different ways members of the human family had adapted to upright walking,[33] long before their brains started expanding in a significant way about 2 million years ago (detailed in Chapter 2 of this book) or before they began using tools about 3 million years ago or so.

While Aiello would tell Darwin that researchers can connect the dots between some of these species,[34] she would also point out that there are gaping holes in the fossil record at other key times—such as the point when one species of *Australopithecus* gave rise to the earliest member of our genus, *Homo*, sometime between 2.5 and 3 million years ago in Africa. This is also during the time when the first stone tools appear in Africa, and researchers who long thought that these tools were made by early members of *Homo* now are considering the possibility that they were invented more than once and, possibly, by a small-brained member of *Australopithecus*.

This is also the time when the first members of the human family apparently trekked out of Africa, where all the action had taken place for the first 4 million years or so of human evolution. But about 2 million years ago, fossils of early hominins appear in Eurasia, notably at the famous site of Dmanisi, Georgia. High on a promontory overlooking the grasslands of the Georgian steppe, researchers have unearthed five skulls and more than fifty bones of the earliest known hominins outside of Africa—members of our genus *Homo* that lived 1.77 million years ago. These people were startlingly primitive, with small bodies about 1.5 meters tall, simple tools, and brains one-third to one-half the size of the brains of modern humans.[35] Yet, they or their ancestors trekked more than 6,000 kilometers from sub-Saharan Africa to the Caucasus Mountains.

Soon after, by 1.6 million years ago, early hominins spread widely over Earth, leaving behind a trail of tools, from North Africa all the way to what is now Java, Indonesia, where a key human ancestor called *Homo erectus* was first discovered. At roughly this time, the brain begins to expand significantly in *H. erectus* in Asia and Africa, and fossil sites show evidence that hominins had joined the carnivores in scavenging and hunting more animals for meat. At Dmanisi, *H. erectus* is found with piles of cobblestones they may have used to stone saber-toothed cats and lion-size hyenas they encountered at the river's edge. Life was undoubtedly dangerous and difficult for these early hominins, but early migrants from Africa eventually gave rise to at least a half dozen different species of archaic hominins that evolved over the past 2 million years or so in Asia and Europe.

By now the burnt rice pudding would be served, and paleoanthropologist Maria Martinón-Torres would join the conversation. As director of the Centro Nacional de Investigación sobre la Evolución Humana in Burgos, Spain, she would tell Darwin firsthand about a twelve-meter-deep pit of bones in her country's Atapuerca Mountains, where some of those descendants have been found, in the form of seventeen ancient skulls of early, or proto, Neanderthals, of the genus *Homo*, dating to about 430,000 years ago. In other nearby sites at Atapuerca, researchers have found the fossils of other species of early *Homo* that date to 600,000–1.1 million years ago. Those species were probably descendants of *Homo erectus*, which either gave rise itself to both modern humans and Neanderthals (*Homo neanderthalensis*)—or to an intermediate species that was the common ancestor of humans and Neanderthals. After proto-Neanderthals split from the lineage leading to modern humans more than 600,000 years ago, Neanderthals evolved in Europe and Asia, where the species persisted until about 35,000 years ago, disappearing soon after modern humans arrived in Europe some 40,000 years ago.

But Neanderthals and modern humans were not alone on the planet; one of the major discoveries of the past decade has been that several different types of humans were alive at the same time over the past half million years, including Denisovans, named for a cave in Siberia where a handful of fossils were found that date to 55,000–250,000 years ago. In 2017, fossils of at least fifteen individuals of a new small-brained species, named *Homo naledi*, found in a pit in the Rising Star Cave in South

Africa, were dated to between 236,000 and 335,000 years. Other types of humans lived on the fringes of Southeast Asia, including a petite human that stood 1.1 meters tall, nicknamed "the Hobbit," which lived on the Indonesian island of Flores sometime between 100,000 and 60,000 years ago. It was found with stone tools and alongside the bones of pygmy elephants (*Stegodon florensis insularis*), which it may have hunted. At roughly the same time, another type of little human, called *Homo luzonensis*, which had tiny molars, lived sometime between 80,000 and 50,000 years ago in a cave on Luzon, the largest island in the Philippines.[36] These ancient hominins on Luzon had curving fingers that suggest they still climbed trees. These diminutive island peoples and the Denisovans may have been remnants of larger populations of descendants of *Homo erectus* that were separated by seas or mountain ranges, so that they evolved in relative isolation, eventually becoming new species or subspecies of humans that adapted to different habitats, just as finches on different islands in the Galápagos evolved into different species.

Darwin might not have been surprised to learn that natural selection favored smaller bodies in humans on islands, just as it does animals—a process known as insular dwarfism, in which animals get smaller on islands where food and large predators are relatively scarce. Interestingly, these diverse groups of hominins all disappeared about the time that our species, *Homo sapiens*, appeared in Asia, says Martinón-Torres, who was part of a team, led by Wu Liu and Xiu-Jie Wu of the Chinese Academy of Sciences' Institute of Vertebrate Paleontology and Paleoanthropology in Beijing, that dated forty-seven teeth of *H. sapiens* from a cave in southern China to 80,000–120,000 years ago.[37] If the dating is accurate, the discovery pushes back the appearance of our species in Asia by at least 30,000 years, challenging the long-standing view that all *H. sapiens* swept out of Africa in a single wave 50,000–70,000 years ago. While genetic studies show that all people outside of Africa are descendants of the same small population of modern humans that migrated out of Africa in the past 70,000 years or so, there is accumulating evidence that other groups of *H. sapiens* left Africa earlier, but their offspring went extinct.

If Darwin's mind wasn't bursting with all this news by now, this would be the time that paleoanthropologist Katerina Harvati of the University of Tübingen in Germany would join the conversation to share with Darwin the fast-moving pace of recent discoveries in the origins and evolution of modern humans. She was part of a team that recently redated a skull from a collapsed cave in Morocco called Jebel Irhoud to a startling 300,000 years ago and concluded that it was the oldest well-dated evidence of an early or putative member of H. sapiens.[38] If it is indeed a proto H. sapiens—and not a late member of another archaic species— that date would push back the appearance of early H. sapiens in the fossil record by about 100,000 years. The oldest widely accepted fossils of H. sapiens are two skullcaps found at Omo Kibish that date to 195,000 years ago and skulls found at Herto that date to 160,000 years ago, both locations in Ethiopia's Great Rift Valley. The Jebel Irhoud people were part of a large, interbreeding population that spread across Africa when the Sahara was green, about 300,000–330,000 years ago; they later evolved as a group toward modern humans.

As noted, some humans then expanded their horizons beyond Africa. Harvati reanalyzed two skulls from Apidima Cave in southern Greece and found that while one was a Neanderthal, as long thought, the other skull looks more like that of an early member of H. sapiens. Her team's new dates place it in Greece as early as 210,000 years ago, which suggests that it (or its ancestors) was part of the wave of modern humans that expanded their range beyond the borders of Africa but went extinct.[39] It is at least 15,000 years older than the next oldest fossil of our species found outside of Africa, from Misliya Cave in Israel, although the dates on both specimens need to be confirmed.[40]

Those new dates fit, however, with the genetic evidence that the ancestors of Homo sapiens split from Neanderthals at least 600,000 years ago. Paleogeneticist Viviane Slon of the Max Planck Institute for Evolutionary Anthropology in Leipzig, Germany, would be the person to tell Darwin about the revolution in genetics that has given researchers a new way to study human evolution. "I would tell Darwin that DNA (the unit of heredity, or what he thought of as 'gemmules') can survive tens and even hundreds of thousands of years after an organism dies, and that we now have techniques to revive and analyze it," says Slon.[41] She would update him on how geneticists have extracted ancient DNA from the bones and teeth of Neanderthals, Denisovans, and other archaic hominins dating back as early as 430,000 years (the proto Neanderthals found at Sima de los Huesos, the pit of bones in Spain's Atapuerca Mountains). By sequencing the entire genome of a pinky finger of a girl living in Denisova Cave more than 55,000 years ago, the Max Planck team, led by paleogeneticist Svante Pääbo, had even discovered a new type of human—the Denisovans—from its DNA, rather than from an analysis of its morphology. They had only a handful of teeth and bone fragments that were too scanty by themselves to identify a new species. Yet, when they analyzed the high-quality genome of the Denisovan girl's pinky finger (and later, DNA from other individuals' teeth), they found they were not Neanderthals, although they were more closely related to Neanderthals than to modern humans. It is a genome in search of a fossil record, and since the discovery, researchers have been hoping to use DNA to find the elusive Denisovans among strange-looking fossils from the past 500,000 years in Asia that have been hard to classify. One candidate was proposed in 2019—a 160,000-year-old fossil jaw found by an unidentified monk in Baishiya Karst Cave in Xiahe county in China at an elevation of 3,200 meters on the margins of the Tibetan Plateau.[42] Paleogeneticists could not get DNA from the jaw to confirm its identity, but they were able to extract collagen, a common structural protein, from one of its large molars. In another demonstration of the power of molecular methods to open new windows on the past, they found its amino acid sequence most closely matched that of Denisovans rather than Neanderthals or Homo sapiens. If this jawbone proves to be a Denisovan, as proposed, it would establish that this enigmatic species ranged far and wide across Asia and that human evolution was far more complex in Asia than previously believed.[43]

Geneticists have also discovered that modern humans outside of Africa repeatedly mated with Neanderthals, overturning old ideas that Neanderthals were ancestral to modern humans or a separate species that could not interbreed with our ancestors. In fact, Chris Stringer might take Darwin on a walk after dinner, perhaps on the Sandwalk, the stone and sand "thinking path" overlooking the Great Pucklands Meadow where Darwin made regular circuits every day at noon. Here, Stringer says, he "would want to update him on 'the wonderful Gibraltar skull' that he saw in 1864, and Neanderthals in general. How our ideas about them have evolved too, including the revelation that Darwin, like all indigenous Europeans, had some Neanderthal-derived DNA in his genome."[44]

The leading hypothesis now is that as a wave of modern humans swept out of Africa in the past 70,000 years or so, they encountered Neanderthals and interbred with them, perhaps in the Middle East. Then, a group of *Homo sapiens* heading into Asia encountered Denisovans, who once ranged widely across Asia, and picked up some of their DNA; today, Europeans and Asians have inherited 1–3 percent of their DNA from their ancestors' encounters with Neanderthals, and indigenous peoples living in Papua New Guinea, Australia, and island Melanesia also carry some Denisovan DNA. Neanderthals and Denisovans also mixed it up, and modern humans who came out of Africa earlier than 70,000 years ago, but went extinct, mated with Neanderthals and with Denisovans.

This ancient DNA from human bones, teeth, and even sediments[45] not only helps researchers sort out how modern humans are related to Neanderthals and other archaic humans, it also gives researchers a view of how natural selection works—and how introgression of genes from Neanderthals and Denisovans can quickly introduce genetic variation into human genomes, as in other mammals. For example, researchers have scanned the genomes of living people in biobanks and found that they have inherited Neanderthal DNA variants that boost some immune responses and are involved in brain development. But many of these variants also may raise the risk of depression, skin lesions, blood clots, and other disorders.[46] Some of those genes may have been optimal for active lives outdoors in the colder climate of prehistoric Europe, rather than in the tropics of Africa, or to battle diseases modern humans hadn't had time to evolve defenses against. By mating with Neanderthals or Denisovans, humans quickly acquired gene variants beneficial to life in Europe or Asia rather than having to wait millennia for natural selection to favor their own versions of genes to survive in these new habitats. But some of those Neanderthal genes that may have been beneficial in prehistoric times are problematic today. People who live indoors more in artificial light, get less exercise, and live to an older age than Neanderthals on average may suffer depression or be prone to strokes—a Neanderthal gene variant that helps blood clot quickly, for example, may have saved prehistoric women hemorrhaging in childbirth or men wounded in fights, but today it increases the risk of stroke in people who live to old age.

Genetic studies also have challenged old views of race, including, as Agustín Fuentes writes in Chapter 7 of this book, Darwin's. Such studies have revealed how little genetic variation there is in modern humans today compared with any two groups of chimpanzees, for example, in Africa. People living outside of Africa, in particular, are closely related to each other—they all are descendants of the single population that migrated out of Africa in the past 70,000 years and underwent a genetic bottleneck that winnowed out much of the genetic variation found in Africans.

In fact, much of the genetic diversity that does exist in people today outside of Africa is the result of their adaptations to new habitats, climates, diseases, and diets. For example, the strongest evidence that natural selection favors genes that help people digest foods that boost fertility can be seen in changes to the lactase gene, which helps humans digest the lactose sugars in milk. In many mammals and humans, this ability disappears after weaning in childhood. In people with lactose tolerance, the gene allows them to digest lactose as adults; and this made them—and their children—more likely to survive and reproduce, thus spreading the gene variant. Similarly, other studies have found that many Greenlandic Inuits have inherited a gene variant that helps them counteract the bad effects of their high-fat diet, based heavily on fish and marine mammals' blubber. And as Holly Dunsworth writes in Chapter 9 of this book, differences in skin color are the result of adaptations to ultraviolet radiation—natural selection favors gene variants for light skin in northern latitudes where there is less ultraviolet radiation and variants for darker skin where there is more sunlight.

As the evening at Down House came to a close, Stringer might reveal one last tantalizing tidbit. He would tell Darwin that all the years he lived in Downe, writing about evolution, he was only thirty kilometers from Swanscombe (he might even have heard of the George & Dragon pub). Nearby, buried in the gravels of the River Thames, was a skull, undiscovered until 1935. Stringer and his colleagues have since identified it as a 400,000-year-old member of *Homo heidelbergensis*, a species that was ancestral to Neanderthals. It was so close—if only Darwin could have seen it and compared it to the Neanderthal Gibraltar skull, he might have seen evidence of evolution in the human family for himself. It might have reminded him of his own prediction that, "*The whole process of that most important function, the reproduction of the species, is strikingly the same in all mammals.*"[47] Including humans.

ACKNOWLEDGMENTS

THIS BOOK WOULD NOT HAVE been possible without the vision, support, and editorial genius of Alison Kalett of Princeton University Press. I hope this is the first of many projects together. Abigail Johnson provided editorial support and helped keep this book on schedule. Thank you to Laurie Schlesinger, David Campbell, Ellen Foos, Kristin Zodrow (in the early days of the project), and many others at Princeton University Press for making this an enjoyable and fulfilling process. I am grateful to Amy K. Hughes for her brilliant copyediting of the original manuscript.

I am awed by my co-contributors. Janet, Alice, Suzana, Brian, Yohannes, Kristina, John, Agustín, Mike, Holly, and Ann: thank you for not only embracing the vision I had for this book but seeing it through beautifully.

Dartmouth College supported this project, in part, through the Fannie and Alan Leslie Dartmouth Conference Fund. I appreciate the encouragement and enthusiastic support I have received throughout my career from the Leakey Foundation, especially Sharal Camisa and Arielle Johnson.

I am grateful to the two anonymous colleagues whose comments made the book proposal for *A Most Interesting Problem* stronger, and the two additional anonymous colleagues who took the time to read and comment on a first draft of this manuscript. The final product benefited from the wisdom of these individuals.

Nate Dominy contributed insightful comments and helpful resources throughout this entire process. Laura Ogden helped me better understand Darwin's ongoing notoriety among the Yagán people of Tierra del Fuego. My students constantly push me to be a better scholar and teacher. In particular, I am grateful to my graduate students (past and present) Alex Claxton, Ellie McNutt, and Kate Miller.

Olivia Fryman, the curator of Down House, graciously welcomed me to Darwin's home. I learned so much there and was inspired by the talented, well-trained, and enthusiastic docents. Jay Satterfield of the Rauner Library at Dartmouth let me read a first-edition *Descent of Man*. Somehow, the words meant a bit more on those original pages.

Permission to use photographs that appear in this book was generously provided by Stephen Atkinson, George Chaplin, José Manuel de la Cuétara, Patrick Evans, Luke Fannin, Aida Gómez-Robles, John Gurche, Nina Jablonski, C.C. Lucas, Gina Martin, Chris Stringer, Vanessa Woods, and the Cambridge University Library.

I am grateful for the many writing and editing tips provided by Neil Young, Lauren Sandler, and Bruce DeSilva. In embarking on this project, I had the opportunity to read many books about the history of science and Darwin's impact on the world. Authors Tasneem Husain, Angela Saini, Maria Popova, David Quammen, Janet Browne, Constance Clark, and Deborah Heligman have written extraordinary books that helped me better formulate the objectives of my own book. I am grateful to Paige Madison and Alex Menez for their expertise in the history of paleoanthropology.

I am forever grateful to Lucy Kirshner and Laura MacLatchy for making me the scientist and science communicator that I am today.

Most of all, I am grateful to my extraordinarily kind, patient, inquisitive, and loving family. Thank you Erin, Ben, and Josie for putting up with my multiyear obsession with Charlie D.

NOTES

Preface

1. Darwin to J. D. Hooker, August 23, 1864, Darwin Correspondence Project, letter no. 4597, https://www.darwinproject.ac.uk/letter/DCP-LETT-4597.xml. In this letter, Darwin wrote in error that he was going to 6 Chester Place.

2. Darwin to Francis Trevelyan Buckland, December 15, 1864, Darwin Correspondence Project, letter no. 4717F, https://www.darwinproject.ac.uk/letter/DCP-LETT-4717F.xml.

3. Alex Menez, "Correspondence: The Day Charles Darwin Came Face to Face with a Neanderthal," *Linnean* 26 (2010): 7.

4. Darwin to J. D. Hooker, September 1, 1864, Darwin Correspondence Project, letter no. 4605, https://www.darwinproject.ac.uk/letter/DCP-LETT-4605.xml.

5. Alex Menez, "The Gibraltar Skull: Early History, 1848–1868," *Archives of Natural History* 45 (2018): 92–110.

6. Francis Darwin, *The Life and Letters of Charles Darwin, Including an Autobiographical Chapter* (London: John Murray, 1887), 1:87.

7. Charles Darwin, *On the Origin of Species by Means of Natural Selection, or the Preservation of Favoured Races in the Struggle for Life*, 1st ed. (London: John Murray, 1859), 482.

8. Ibid., 490.

9. Ibid., 488.

10. Darwin to Alfred Russel Wallace, December 22, 1857, Darwin Correspondence Project, letter no. 2192, https://www.darwinproject.ac.uk/letter/DCP-LETT-2192.xml.

11. Charles Darwin, *The Descent of Man, and Selection in Relation to Sex*, 1st ed. (London: John Murray, 1871).

12. Janet E. Browne, *Darwin: The Power of Place* (New York: Knopf, 2002), 8.

Introduction

Some of this introduction draws on material previously published in *The Descent of Man* (Ware, UK: Wordsworth Editions, 2013). I gratefully acknowledge permission to use this material.

1. Charles Darwin, *The Descent of Man, and Selection in Relation to Sex*, 1st ed. (London: John Murray, 1871), 2:405.

2. Charles Darwin, *On the Origin of Species by Means of Natural Selection, or the Preservation of Favoured Races in the Struggle for Life*, 1st ed. (London: John Murray, 1859), 488.

3. Peter Bowler, *Monkey Trials and Gorilla Sermons: Evolution and Christianity from Darwin to Intelligent Design* (Cambridge, MA: Harvard University Press, 2007); David N. Livingstone, *Darwin's Forgotten Defenders: The Encounter between Evangelical Theology and Evolutionary Thought*, 2nd ed. (Vancouver, BC: Regent College Pub., 2001). See also: Janet E. Browne, *Darwin: Voyaging* (New York: Knopf, 1995), and *Darwin: The Power of Place* (New York: Knopf, 2002); James A. Secord, ed. *Charles Darwin: Evolutionary Writings* (Oxford: Oxford University Press, 2008).

4. Adrian Desmond, *The Politics of Evolution: Morphology, Medicine, and Reform in Radical London* (Chicago: University of Chicago Press, 1989).

5. Darwin, *Descent*, 1:2. "Of the older and honoured chiefs in natural science, many unfortunately are still opposed to evolution in any form."

6. There are several claims about Darwin's first use of this phrase. In 1868, Alfred Russel Wallace urged him to adopt Spencer's phrase as an alternative to "natural selection." Darwin consequently used it here and there in *The Variation of Animals and Plants under Domestication* (London: John Murray, 1868) and in the 5th ed. of *Origin* (1869); see Browne, *Power of Place*, 312–14. Darwin refers to it in *Descent*, 1:152.

7. Of the many scholarly studies of Darwin's life and times, see especially Peter Bowler, *Evolution: The History of An Idea* (Berkeley: University of California Press, 1984); Jonathan Hodge and Gregory Radick, eds., *The Cambridge Companion to Darwin*, 2nd ed. (Cambridge: Cambridge University Press, 2009); and Darwin's own recollections published in Nora Barlow, ed., *The Autobiography of Charles Darwin, 1809–1882; With the Original Omissions Restored* (London: Collins, 1958). Two websites are also invaluable resources:

Darwin Correspondence Project: http://www.darwinproject.ac.uk/

The Complete Work of Charles Darwin Online, John van Wyhe, ed. (2002–): http://darwin-online.org.uk/

8. Jenny Uglow, *The Lunar Men: Five Friends Whose Curiosity Changed the World* (New York: Farrar, Straus and Giroux, 2003).

9. Aileen Fyfe and Bernard Lightman, eds., *Science in the Marketplace: Nineteenth-Century Sites and Experiences* (Chicago: University of Chicago Press, 2007).

10. Darwin to Wallace, December 22, 1857, Frederick Burkhardt and Sydney Smith, eds., *The Correspondence of Charles Darwin*, 26 vols. (Cambridge: Cambridge University Press, 1985–2019), vol. 6:515.

11. James A. Secord, *Victorian Sensation: The Extraordinary Publication, Reception, and Secret Authorship of Vestiges of the Natural History of Creation* (Chicago: University of Chicago Press, 2000).

12. Roger Cooter, *The Cultural Meaning of Popular Science: Phrenology and the Organization of Consent in Nineteenth-Century Britain* (Cambridge: Cambridge University Press, 1984).

13. Charles Darwin, *Journal of Researches into the Geology and Natural History of the Various Countries Visited by H.M.S. Beagle* (London, 1839), 228.

14. A. J. Desmond and J. R. Moore, *Darwin's Sacred Cause: Race, Slavery, and the Quest for Human Origins* (London, New York: Allen Lane, 2009).

15. Ian Hesketh, *Of Apes and Ancestors: Evolution, Christianity, and the Oxford Debate* (Toronto: University of Toronto Press, 2009).

16. A full account of Huxley's life in science is given in A. J. Desmond, *Huxley*, 2 vols. (London: Michael Joseph, 1994, 1997).

17. Browne, *Power of Place*, 222.

18. Darwin to Wallace, March 27, 1869, Burkhardt and Smith, *Correspondence of Charles Darwin*, 17:157.

19. Browne, *Power of Place*, 74–76.

20. Darwin, *Descent*, 1:35.

21. Frances Power Cobbe, *Darwinism in Morals and Other Essays* (London: Williams and Norgate, 1872).

22. Stephen G. Alter, *Darwinism and the Linguistic Image: Language, Race, and Natural Theology in the Nineteenth Century* (Baltimore: Johns Hopkins University Press, 1999).

23. Darwin, *Descent*, 1:68–69.

24. Ibid., 73.

25. Ibid., 101.

26. Ibid.

27. Ibid., 70–71.

28. Darwin, *Descent*, 2:404–5.

29. Darwin, *Descent*, 1:206–7.

30. Ibid., 249.

31. Darwin to Wallace, February 26, 1867, Francis Darwin, *The Life and Letters of Charles Darwin, Including an Autobiographical Chapter* (London: John Murray, 1887), 3:95.

32. Ibid., 91.

33. Helena Cronin, *The Ant and the Peacock: Altruism and Sexual Selection from Darwin to Today* (Cambridge: University of Cambridge Press, 1991); Bernard Campbell, ed., *Sexual Selection and the Descent of Man, 1871–1971* (Chicago: Aldine Publishing, 1972). The most recent and important study is by Evelleen Richards, *Darwin and the Making of Sexual Selection* (Chicago: University of Chicago Press, 2017).

34. Darwin, *Descent*, 2:368–69.

35. Kimberly Hamlin, *From Eve to Evolution: Darwin, Science and Women's Rights in Gilded Age America* (Chicago: University of Chicago Press, 2014).

36. Darwin, *Descent*, 1:160.

37. Social Darwinism is discussed in many texts. See Mike Hawkins, *Social Darwinism in European and American Thought, 1860–1945: Nature as Model and Nature as Threat* (Cambridge: Cambridge University Press, 1997).

38. Karl Pearson, *The Grammar of Science* (London: Adam and Charles Black, 1900), 369.

Chapter 1: The Fetus, the Fish Heart, and the Fruit Fly

1. Charles Darwin, *The Descent of Man, and Selection in Relation to Sex*, 1st ed. (London: John Murray, 1871), 1:2.

2. Ibid., 14.

3. Charles Darwin, *On the Origin of Species by Means of Natural Selection, or the Preservation of Favoured Races in the Struggle for Life*, 1st ed. (London: John Murray, 1859), 368.

4. Walid A. Zaher et al., "Sternalis: A Clinically Important Variation," *Pakistan Journal of Medical Science* 25 (2009): 325–28.

5. Sandeep J. Sebastin et al., "Does the Absence of the Palmaris Longus Affect Grip and Pinch Strength?" *Journal of Hand Surgery* 30B (2005): 406–8.

6. Darwin, *Descent*, 1:21.

7. Tiffany Y. Loh and Philip R. Cohen, "Darwin's Tubercle: Review of a Unique Congenital Anomaly," *Dermatology and Therapy* 6 (2016): 143–49.

8. John P. McGann, "Poor Human Olfaction is a Nineteenth Century Myth," *Science* 356 (2017): eaam7263.

9. Darwin, *Descent*, 1:24–25.

10. R. Randal Bollinger et al., "Biofilms in the Large Bowel Suggest an Apparent Function of the Human Vermiform Appendix," *Journal of Theoretical Biology* 249 (2007): 826–31.

11. Jeffrey Lordan, Peter Rauh, and Robert J. Spinner, "The Clinical Anatomy of the Supracondylar Spur and the Ligament of Struthers," *Clinical Anatomy* 18 (2005): 548–51.

12. Waseem Bhat et al., "Painful, Palpable and Pathological: Anomalous Flexor Digitorum Superficialis Brevis in the Palm, Comparative Anatomical Context, and an Updated Classification of Anomalies of the Flexor Digitorum Superficialis," *Journal of Hand Surgery* 39E(1) (2014): 101–6.

13. Paul Z. Myers, "*Hox* Genes in Development: the *Hox* Code," *Nature Education* 1 (2008): 2.

14. Darwin, *Descent*, 1:31.

15. Richard Owen, *On the Nature of Limbs: A Discourse Delivered on Friday, February 9, at an Evening Meeting of the Royal Institution of Great Britain* (London: John van Voorst, 1849).

16. Darwin, *Descent*, 1:32.

17. Darwin, *Origin*, 479.

18. Etienne G. J. Danchin and Pierre Pontarotti, "Statistical Evidence for a More Than 800-Million-Year-Old Evolutionarily Conserved Genomic Region in Our Genome," *Journal of Molecular Biology and Evolution* 59 (2004): 587–97.

19. Alice M. Roberts, *The Incredible Unlikeliness of Being: Evolution and the Making of Us* (London: Bloomsbury, 2015).

Chapter 2: Remarkable but Not Extraordinary:
The Evolution of the Human Brain

1. Charles Darwin, *The Descent of Man, and Selection in Relation to Sex*, 1st ed. (London: John Murray, 1871), 1:35.

2. Charles G. Gross, *Brain, Vision, Memory: Tales in the History of Neuroscience* (Cambridge, MA: MIT Press, 1998).

3. Darwin, *Descent*, 1:34.

4. Ibid., 35. He uses the term "lower animals" eighty-one times in the 1st ed. of vol. 1 of *Descent*.

5. Ibid., 37.

6. Marshall Hall, "On the Reflex Function of the Medulla Oblongata and Medulla Spinalis," *Philosophical Transactions* 123 (1833): 625–65.

7. Charles S. Sherrington, *The Integrative Action of the Nervous System* (New Haven, CT: Yale University Press, 1906).

8. Sten Grillner and Peter Wallén, "Central Pattern Generators for Locomotion, with Special Reference to Vertebrates," *Annual Review of Neuroscience* 8 (1985): 233–61.

9. Shy Shoham, Daniel H. O'Connor, and Ronen Segev, "How Silent Is the Brain: Is There a 'Dark Matter' Problem in Neuroscience?" *Journal of Comparative Physiology A* 192 (2006): 777–84.

10. Rodrigo Q. Quiroga et al., "Invariant Visual Representation by Single Neurons in the Human Brain," *Nature* 435 (2005): 1102–7; Stephen M. Kosslyn, Giorgio Ganis, and William L. Thompson, "Neural Foundations of Imagery," *Nature Reviews Neuroscience* 2 (2001): 635–42; William E. Skaggs and Bruce L. McNaughton, "Replay of Neuronal Firing Sequences in Rat Hippocampus during Sleep Following Spatial Experience," *Science* 271 (1996): 1870–73.

11. Murray Shanahan et al., "Large-Scale Network Organization in the Avian Forebrain: A Connectivity Matrix and Theoretical Analysis," *Frontiers in Computational Neuroscience* 7 (2013): 89.

12. Dante Mantini et al., "Default Mode of Brain Function in Monkeys," *Journal of Neuroscience* 31 (2011): 12954–62; Hanbing Lu et al., "Rat Brains Also Have a Default Mode Network," *Proceedings of the National Academy of Sciences* 109 (2012): 3979–84; Marcus E. Raichle et al., "A Default Mode of Brain Function," *Proceedings of the National Academy of Sciences* 98 (2001): 676–82; Linda J. Larson-Prior et al., "Modulation of the Brain's Functional Network Architecture in the Transition from Wake to Sleep," *Progress in Brain Research* 193 (2011): 277.

13. Fabio Ferrarelli et al., "Breakdown in Cortical Effective Connectivity during Midazolam-Induced Loss of Consciousness," *Proceedings of the National Academy of Sciences* 107 (2010): 2681–86.

14. Xiaoning Han et al., "Forebrain Engraftment by Human Glial Progenitor Cells Enhances Synaptic Plasticity and Learning in Adult Mice," *Cell Stem Cell* 12 (2013): 342–53.

15. Bosiljka Tasic et al., "Adult Mouse Cortical Cell Taxonomy Revealed by Single Cell Transcriptomics," *Nature Neuroscience* 19 (2016): 335–46; Spyros Darmanis et al., "A Survey of Human Brain Transcriptome Diversity at the Single Cell Level," *Proceedings of the National Academy of Sciences* 112 (2015): 7285–90.

16. Reviewed in Suzana Herculano-Houzel, "Humans rule!," in *The Human Advantage: How Our Brains Became Remarkable* (Cambridge: MIT Press, 2016), chap. 1.

17. Suzana Herculano-Houzel, *The Human Advantage: How Our Brains Became Remarkable* (Cambridge: MIT Press, 2016).

18. Bruno Mota and Suzana Herculano-Houzel, "Cortical Folding Scales Universally with Cortical Surface Area and Thickness, Not Number of Neurons," *Science* 349 (2015): 74–77.

19. Suzana Herculano-Houzel, "Scaling of Brain Metabolism with a Fixed Energy Budget Per Neuron: Implications for Neuronal Activity, Plasticity and Evolution," *PLoS One* 6 (2011): e17514.

20. Karina Fonseca-Azevedo and Suzana Herculano-Houzel, "Metabolic Constraint Imposes Trade-off between Body Size and Number of Brain Neurons in Human Evolution," *Proceedings of the National Academy of Sciences* 109 (2012): 18571–76.

21. Suzana Herculano-Houzel, "Neuronal Scaling Rules for Primate Brains: The Primate Advantage," *Progress in Brain Research* 195 (2012): 325–40.

22. Mariana Gabi et al., "No Expansion in Numbers of Prefrontal Neurons in Primate and Human Evolution," *Proceedings of the National Academy of Sciences* 113 (2016): 9617–22.

23. Alex Wissner-Gross, "A New Equation for Intelligence," TEDxBeaconStreet, November 2013, https://www.ted.com/talks/alex_wissner_gross_a_new_equation_for_intelligence?language=en.

24. Suzana Herculano-Houzel, "Longevity and Sexual Maturity Vary across Species with Number of Cortical Neurons, and Humans Are No Exception," *Journal of Comparative Neurology* 527 (2019): 1689–1705.

25. Suzana Herculano-Houzel, "Life History Changes Accompany Increased Numbers of Cortical Neurons: A New Framework for Understanding Human Brain Evolution," *Progress in Brain Research* 250 (2019): 179–216.

Chapter 3: The Darwinian Road to Morality

1. Charles Darwin, *The Descent of Man, and Selection in Relation to Sex*, 1st ed. (London: John Murray, 1871), 1:71–72.

2. Janet Browne, *Darwin: Voyaging* (Princeton, NJ: Princeton University Press, 1996).

3. Stefan Kühl, *For the Betterment of the Race: The Rise and Fall of the International Movement for Eugenics and Racial Hygiene* (New York: Palgrave Macmillan, 2013).

4. Darwin, *Descent*, 1:82.

5. Brian Hare and Vanessa Woods, *Survival of the Friendliest* (New York: Penguin Random House, 2020).

6. Martin N. Muller, Richard W. Wrangham, and David R. Pilbeam, eds., *Chimpanzees and Human Evolution* (Cambridge, MA: Harvard University Press, 2017).

7. Brian Hare and Shinya Yamamoto, *Bonobos: Unique in Mind, Brain, and Behavior* (Oxford: Oxford University Press, 2017).

8. Darwin, *Descent*, 1:77.

9. Hare and Yamamoto, *Bonobo*.

10. Michael Tomasello, *Becoming Human: A Theory of Ontogeny* (Cambridge, MA: Belknap Press, 2019).

11. Hare and Woods, *Survival of the Friendliest*.

12. Ibid.

13. Darwin, *Descent*, 1:91.

14. Gema Martin-Ordas et al., "Keeping Track of Time: Evidence for Episodic-like Memory in Great Apes," *Animal Cognition* 13, no. 2 (2010): 331–40.

15. Alexandra G. Rosati and Brian Hare, "Chimpanzees and Bonobos Exhibit Emotional Responses to Decision Outcomes," *PloS One* 8, no. 5 (2013): e63058.

16. Darwin, *Descent*, 1:72.

17. Jane Goodall, "Learning from the Chimpanzees: A Message Humans Can Understand," *Science* 282 (1998): 2184–85.

18. Thibaud Gruber, Zanna Clay, and Klaus Zuberbühler, "A Comparison of Bonobo and Chimpanzee Tool Use: Evidence for a Female Bias in the *Pan* Lineage," *Animal Behaviour* 80, no. 6 (2010): 1023–33.

19. Andrew Whiten, Victoria Horner, and Frans B. de Waal, "Conformity to Cultural Norms of Tool Use in Chimpanzees," *Nature* 437, no. 7059 (2005): 737.

20. Darwin, *Descent*, 1:105.

21. Brian Hare, "Survival of the Friendliest: *Homo sapiens* Evolved via Selection for Prosociality," *Annual Review of Psychology* 68 (2017): 155–86.

22. Brian Hare and Vanessa Woods, *The Genius of Dogs* (London: Oneworld Publications, 2012).

23. Darwin, *Descent*, 1:77.

24. Ibid., 80.

25. Hare and Woods, *Survival of the Friendliest.*

26. Darwin, *Descent*, 1:91.

27. Hare and Woods, *Survival of the Friendliest.*

28. Ibid.

29. Ibid.

30. Alfred Tennyson, *In Memoriam AHH* (Los Angeles: Hardpress Publishing, 2012 [1850]).

31. Hare and Woods, *Survival of the Friendliest.*

Chapter 4: Charles Darwin and the Fossil Evidence for Human Evolution

1. Charles Darwin, *The Descent of Man, and Selection in Relation to Sex*, 1st ed. (London: John Murray, 1871), 1:144–45.

2. Charles Darwin, *On the Origin of Species by Means of Natural Selection, or the Preservation of Favoured Races in the Struggle for Life*, 1st ed. (London: John Murray, 1859).

3. Charles Darwin, *The Descent of Man and Selection in Relation to Sex*, 2nd ed. (London: John Murray, 1874).

4. Darwin, *Descent*, 1st ed., 1:199.

5. Raymond Dart, "*Australopithecus africanus*: The Man-Ape of South Africa," *Nature* 115 (1925): 195–99.

6. Darwin, *Descent*, 1st ed., 1:141.

7. Yohannes Haile-Selassie et al., "Preliminary Geology and Paleontology of New Hominid-Bearing Pliocene Localities in the Central Afar Region of Ethiopia," *Anthropological Science* 115(3) (2007): 215–22.

8. Michel Brunet et al., "A New Hominid from the Upper Miocene of Chad, Central Africa," *Nature* 418 (2002): 145–51.

9. Brigitte Senut et al., "First Hominid from the Miocene (Lukeino Formation, Kenya)," *Comptes Rendus de l'Académie des Sciences*, ser. 2A, *Earth and Planetary Science* 332 (2001): 137–44.

10. Yohannes Haile-Selassie, "Late Miocene Hominids from the Middle Awash, Ethiopia," *Nature* 412 (2001): 178–81.

11. Tim D. White, Gen Suwa, and Berhane Asfaw, "*Australopithecus ramidus*, a New Species of Early Hominid from Aramis, Ethiopia," *Nature* 371 (1994): 306–12.

12. Tim D. White et al., "*Ardipithecus ramidus* and the Paleobiology of Early Hominids," *Science* 326 (2009): 64–86.

13. Darwin, *Descent*, 1st ed., 1:199

14. T. D. White, C. O. Lovejoy, B. Asfaw, J. P. Carlson, and G. Suwa, "Neither Chimpanzee nor Human, *Ardipithecus* Reveals the Surprising Ancestry of Both," *Proceedings of the National Academy of Sciences* 112(16) (2015): 4877–84.

15. Darwin, *Descent*, 1st ed., 1:142.

16. Meave G. Leakey, Craig S. Feibel, Ian McDougall, and Alan Walker, "New Four-Million-Year-Old Hominid Species from Kanapoi and Allia Bay, Kenya," *Nature* 376 (1995): 565–71.

17. Tim D. White et al., "Asa Issie, Aramis and the Origin of *Australopithecus*," *Nature* 440 (2006): 883–89.

18. Yohannes Haile-Selassie et al., "A 3.8-Million-Year-Old Hominin Cranium from Woranso-Mille, Ethiopia," *Nature* 573 (2019): 214–19.

19. Donald C. Johanson and Maitland Edey, *Lucy: The Beginnings of Humankind* (New York: Simon & Schuster, 1981).

20. Donald C. Johanson, "Lucy, Thirty Years Later: An Expanded View of *Australopithecus afarensis*," *Journal of Anthropological Research* 60(4) (2004): 465–86.

21. Zeresenay Alemseged et al., "A Juvenile Early Hominin from Dikika, Ethiopia," *Nature* 443 (2006): 296–301.

22. Yohannes Haile-Selassie et al., "An Early *Australopithecus afarensis* Postcranium from Woranso-Mille, Ethiopia," *Proceedings of the National Academy of Sciences* 107 (2010): 12121–26.

23. Darwin, *Descent*, 1st ed., 1:143.

24. Yohannes Haile-Selassie et al., "New Species from Ethiopia Further Expands Middle Pliocene Hominin Diversity," *Nature* 521 (2015): 483–88.

25. Yohannes Haile-Selassie et al., "A New Hominin Foot from Ethiopia Shows Multiple Pliocene Bipedal Adaptations," *Nature* 483 (2012): 565–69.

26. Mary D. Leakey and Richard L. Hay, "Pliocene Footprints in the Laetolil Beds at Laetoli, Northern Tanzania," *Nature* 278 (1979): 317–23.

27. Yohannes Haile-Selassie, Stephanie M. Melillo, and Denise Su, "The Pliocene Hominin Diversity Conundrum: Do More Fossils Mean Less Clarity?" *Proceedings of the National Academy of Sciences* 113(23) (2016): 6364–71.

28. Meave G. Leakey et al., "New Hominin Genus from Eastern Africa Shows Diverse Middle Pliocene Lineages," *Nature* 410 (2001): 433–40

29. Darwin, *Descent*, 1st ed., 1:226–27.

30. Shannon P. McPherron et al., "Evidence for Stone-Tool-Assisted Consumption of Animal Tissues before 3.39 Million Years Ago at Dikika, Ethiopia," *Nature* 466 (2010): 857–60.

31. Sonia Harmand et al., "3.3-Million-Year-Old Stone Tools from Lomekwi 3, West Turkana, Kenya," *Nature* 521 (2015): 310–15.

32. Berhane Asfaw et al., "*Australopithecus garhi*: A New Species of Early Hominid from Ethiopia," *Science* 284 (1999): 629–35.

33. Sileshi Semaw et al., "2.5-Million-Year-Old Stone Tools from Gona, Ethiopia," *Nature* 385 (1997): 333–36.

34. Ronald J. Clarke and Kathleen Kuman, "The Skull of StW 573, a 3.67 Ma *Australopithecus prometheus* Skeleton from Sterkfontein Caves, South Africa," *Journal of Human Evolution* 134 (2019): 102634.

35. Lee R. Berger et al., "*Australopithecus sediba*: A New Species of *Homo*-Like Australopith from South Africa," *Science* 328 (2010): 195–204.

36. Andy Herries et al., "Contemporaneity of *Australopithecus, Paranthropus*, and early *Homo erectus* in South Africa," *Science* 368 (2020), eaaw7293.

37. Bernard Wood and Kes Schroer, "*Paranthropus*," in *Companion to Paleoanthropology*, ed. David Begun (New York: Wiley-Blackwell, 2013), 457–78.

38. Brian Villmoare et al., "Early *Homo* at 2.8 Ma from Ledi-Geraru, Afar, Ethiopia," *Science* 347 (2015): 1352–55; William H. Kimbel, Donald C. Johanson, and Yoel Rak, "Systematic Assessment of a Maxilla of *Homo* from Hadar, Ethiopia," *American Journal of Physical Anthropology* 103(2) (1997): 235–62.

39. Louis S. B. Leakey, Phillip V. Tobias, and John R. Napier, "A New Species of the Genus *Homo* from Olduvai Gorge," *Nature* 202 (1964): 7–9.

40. Bernard A. Wood, "Origin and Evolution of the Genus *Homo*," *Nature* 355 (1992): 783–90.

41. Herries et al., 2020.

42. David Lordkipanidze et al., "A Complete Skull from Dmanisi, Georgia, and the Evolutionary Biology of Early *Homo*," *Science* 342 (2013): 326–31.

43. Darwin, *Descent*, 1st ed., 1:137

44. William King, "The Reputed Fossil Man of the Neanderthal," *Quarterly Journal of Science* 1 (1864), 88–97.

45. Lee R. Berger et al., "*Homo naledi*, a New Species of the Genus *Homo* from the Dinaldi Chamber, South Africa," *eLife* (2015): e09560, DOI:10.7554/eLife.09560; Michael J. Morwood et al., "Archaeology and Age of a New Hominin from Flores in Eastern Indonesia," *Nature* 432 (2004): 1087–91; Florent Détroit et al., "A New Species of *Homo* from the Late Pleistocene of Philippines," *Nature* 568 (2019): 181–86.

46. Johannes Krause et al., "The Complete Mitochondrial DNA Genome of an Unknown Hominin from Southern Siberia," *Nature* 464 (2010): 894–97.

47. Tim D. White et al., "Pleistocene *Homo sapiens* from Middle Awash, Ethiopia," *Nature* 423 (2003): 742–47; Jean-Jacques Hublin et al., "New Fossils from Jebel Irhoud, Morocco and the Pan-African Origin of *Homo sapiens*," *Nature* 546 (2017): 289–92.

48. Thomas H. Huxley, *Evidence as to Man's Place in Nature* (London: Williams and Norgate, 1863).

49. Darwin, *Descent*, 1st ed., 1:199.

Chapter 5: A Century of Civilization, Intelligence, and (White) Nationalism

1. Charles Darwin, *The Descent of Man, and Selection in Relation to Sex*, 1st ed. (London: John Murray, 1871), 1:158.

2. Jon Marks, "The Coevolution of Human Origins, Human Variation, and Their Meaning in the 19th Century," *Zygon* 54, no. 1 (2019): 246–51; Tim Murray, *From Antiquarian to Archaeologist: The History and Philosophy of Archaeology* (South Yorkshire: Pen and Sword, 2014); Alfred R. Wallace, "Anthropology: Address," *Report of the British Association for the Advancement*

of Science 36th Meeting (Nottingham, 1866), 93–94. In 1866, Wallace was one of the first to define a holistic anthropology, in his address to the British Association for the Advancement of Science: "Anthropology is the science which contemplates man under all his various aspects (as an animal, and as a moral and intellectual being) in his relations to lower organisms, to his fellow men, and to the universe. The anthropologist seeks to collect together and systematize the facts and the laws which have been brought to light by all those branches of study which, directly or indirectly, have man for their object. These are very various."

3. Marks, "Coevolution of Human Origins," 246–51, quote on 248.

4. Jeremy Vetter, "The Unmaking of an Anthropologist: Wallace Returns from the Field, 1862–70," *Notes and Records of the Royal Society* 64 (2009): 25–42.

5. Ibid.

6. Alfred R. Wallace, "The Origin of Human Races and the Antiquity of Man Deduced from the Theory of 'Natural Selection,'" *Anthropological Review* (May 1864): clxii.

7. Ibid.

8. Ibid., clxii–clxiv.

9. Ibid., clxiv.

10. Ibid., clxiv–clxv.

11. Ibid., clxv.

12. Ibid., clxiv–clx.

13. Vetter, "Unmaking of an Anthropologist," 29.

14. Darwin, *Descent*, 1:157.

15. Ibid., 159.

16. Ibid., 158.

17. Ibid., 160.

18. Ibid., 162.

19. Ibid., 162–63.

20. Ibid., 163.

21. Ibid., 165.

22. Ibid., 165–66.

23. Ibid., 166.

24. Ibid., 167.

25. Ibid., 168.

26. Ibid., 168–69.

27. Randal Keynes, *Darwin, His Daughter, and Human Evolution* (New York: Penguin Random House, 2002); Lukas Fenner, Matthias Egger, and Sebastien Gagneux, "Annie Darwin's Death, the Evolution of Tuberculosis, and the Need for Systems Epidemiology," *International Journal of Epidemiology* 38, no. 6 (2009): 1425–28.

28. Darwin, *Descent*, 1:174.

29. Ibid., 172.

30. Ibid., 172–74.

31. Ibid., 177.

32. Ibid.

33. Ibid., 178.

34. Ibid.

35. Ibid., 179.

36. Ibid.

37. Ibid., 183.

38. Lewis H. Morgan, *Ancient Society; or, Researches in the Lines of Human Progress from Savagery, through Barbarism to Civilization* (New York: H. Holt, 1877).

39. Darwin, *Descent*, 1:184.

40. Marks, "Coevolution of Human Origins," 250.

41. Darwin, *Descent*, 1:145–46.

42. Henry H. Goddard, "Mental Tests and the Immigrant," *Journal of Delinquency* 2, no. 5 (1917): 243–77.

43. Stephen J. Gould, *The Mismeasure of Man* (New York: W. W. Norton, 1981; reprinted and expanded in 1996), 52.

44. Gould, *Mismeasure*, 28.

45. Eric Siegel, "The Real Problem with Charles Murray and 'The Bell Curve,'" *Scientific American* (blog), April 12, 2017, https://blogs.scientificamerican.com/voices/the-real-problem-with-charles-murray-and-the-bell-curve/.

46. Howard Gardner, "Cracking Open the IQ Box," *American Prospect*, December 10, 2001, https://prospect.org/article/cracking-open-iq-box.

47. Darwin, *Descent*, vol. 1.

48. Charlotte Hunt-Grubbe, "The Elementary DNA of Dr. Watson," *Sunday Times*, October 14, 2007, https://www.thetimes.co.uk/article/the-elementary-dna-of-dr-watson-gllb6w2vpdr.

49. Gould, *Mismeasure*, 28.

50. Trip Gabriel, "A Timeline of Steve King's Racist Remarks and Divisive Actions," *New York Times*, January 15, 2019, https://www.nytimes.com/2019/01/15/us/politics/steve-king-offensive-quotes.html.

51. David S. Shields, "Civilization," in *Keywords for American Cultural Studies*, 2014, https://keywords.nyupress.org/american-cultural-studies/essay/civilization/.

52. Ralph W. Emerson, "American Civilization," *Atlantic* (April 1862): 502–511, https://www.theatlantic.com/magazine/archive/1862/04/american-civilization/306548/.

53. Emile Durkheim and Marcel Mauss, "Note on the Notion of Civilization" (1913), translated by Benjamin Nelson, *Social Research*, 38, no. 4 (1971): 808–13.

54. Julian S. Huxley, "Guest Editorial: Evolution, Cultural and Biological," *Yearbook of Anthropology* (1955): 2–25, quote on 15.

55. Huxley, "Evolution, Cultural and Biological," 17.

56. Darwin, *Descent*, 1:167.

57. Jared Diamond, *Collapse: How Societies Choose to Fail or Succeed* (New York: Viking Press, 2005); Yuval Noah Harari, *Sapiens: A Brief History of Humankind* (New York: Harper, 2015).

58. See, for example, Patricia A. McAnany and Norman Yoffee, eds., *Questioning Collapse: Human Resilience, Ecological Vulnerability, and the Aftermath of Empire* (Cambridge: Cambridge University Press, 2009); Jane E. Buikstra, *Bioarchaeologists Speak Out: Deep Time Perspectives on Contemporary Issues* (n.p., Switzerland: Springer, 2019); Nicola Terrenato, *The Early Roman*

Expansion into Italy: Elite Negotiation and Family Agendas (Cambridge: Cambridge University Press, 2019).

59. Kent V. Flannery, "The Cultural Evolution of Civilizations," *Annual Review of Ecology and Systematics* 3, no.1 (1972): 399–426; Lewis R. Binford, *Mortuary Practices: Their Study and Their Potential: An Archaeological Perspective* (London: Seminar Press, 1972); Michael B. Schiffer, "Archaeological Context and Systemic Context," *American Antiquity* 37, no. 2 (1972): 156–65.

60. Robert Chapman, *Archaeologies of Complexity* (New York: Routledge, 2003), 7.

61. Ibid., 195.

62. David Standish, "Kurt Vonnegut: Playboy Interview (1973)," *Scraps from the Loft*, October 4, 2016, https://scrapsfromtheloft.com/2016/10/04/kurt-vonnegut-playboy-interview/.

Chapter 6: Ranking Humanity among the Primates

1. Charles Darwin, *The Descent of Man, and Selection in Relation to Sex*, 1st ed. (London: John Murray, 1871), 1:191.

2. Charles Lyell, *The Geological Evidences of the Antiquity of Man: With Remarks on Theories of the Origin of Species by Variation* (London: John Murray, 1863); Thomas Henry Huxley, *Evidence as to Man's Place in Nature* (London: Williams and Norgate, 1863); St. George Mivart, "8. Contributions towards a More Complete Knowledge of the Axial Skeleton in Primates," *Proceedings of the Zoological Society of London* 33, no. 1 (1865): 545–92; St. George Mivart, "XIII. On the Appendicular Skeleton of the Primates," *Philosophical Transactions of the Royal Society of London* 157 (1867): 299–429.

3. Darwin, *Descent*, 1:201.

4. J. David Archibald, *Aristotle's Ladder, Darwin's Tree: The Evolution of Visual Metaphors for Biological Order* (New York: Columbia University Press, 2014).

5. Édouard Lartet, "Note Sur un Grand Singe Fossile Qui se Rattache au Groupe des Singes Superieurs." *Comptes Rendus Hebdomadaires des Seances de l'Academie des Sciences* 43 (1856): 219–23.

6. Ernst Haeckel, *Generelle Morphologie der Organismen: Allgemeine Grundzüge der organischen Formen-Wissenschaft, mechanisch begründet durch die von C. Darwin reformirte Descendenz-Theorie, etc.* (Berlin: Georg Reimer Verlag, 1866).

7. Darwin, *Descent*, 1:199.

8. Ibid., 200.

9. Albert Gaudry, *Animaux Fossiles et Géologie de l'Attique: d'Après les Recherches Faites en 1855–56 et en 1860 Sous les Auspices de l'Académie des Sciences*, vol. 1 (Paris: Savy, 1862).

10. Knut Finstermeier et al., "A Mitogenomic Phylogeny of Living Primates," *PLoS One* 8, no. 7 (2013): e69504.

11. Matt Cartmill, "Assessing Tarsier Affinities: Is Anatomical Description Phylogenetically Neutral?" *Geobios* 15 (1982): 279–87.

12. Carl von Linnaeus, *Systema Naturae*, vol. 1 (Holmiae: Salvius, 1758); Étienne Geoffroy Saint-Hilaire, *Tableau des Quadrumanes, ou des Animaux Composant le Premier Ordre de la Classe des Mammifères* (Paris: Muséum National d'Histoire Naturelle, 1812).

13. Mivart, "Axial Skeleton in Primates"; Ernst Haeckel, *Generelle Morphologie*.

14. Reginald I. Pocock, "On the External Characters of the Lemurs and of Tarsius," *Proceedings of the Zoological Society of London* 88, no. 1–2 (1918): 19–53.

15. Christopher Beard, *The Hunt for the Dawn Monkey: Unearthing the Origins of Monkeys, Apes, and Humans* (Berkeley: University of California Press, 2004).

16. Natalie M. Jameson et al., "Genomic Data Reject the Hypothesis of a Prosimian Primate Clade," *Journal of Human Evolution* 61, no. 3 (2011): 295–305.

17. Darwin, *Descent*, 1:200.

18. Ibid., 199.

19. Ibid.

20. Johann F. Blumenbach, *De Generis Humani Variate Nativa* (Göttingen: University of Göttingen, 1775).

21. Darwin, *Descent*, 1:199.

22. Ibid.

23. Raymond A. Dart, "*Australopithecus africanus*: The Man-Ape of South Africa," *Nature* 115 (1925): 195–99; Louis S. B. Leakey, "Africa's Contribution to the Evolution of Man," *South African Archaeological Bulletin* 16, no. 61 (1961): 3–7; Tim D. White et al., "*Ardipithecus ramidus* and the Paleobiology of Early Hominids," *Science* 326, no. 5949 (2009): 64–86.

24. "Celebrate Africa Day, Celebrate Humanity," Maropeng and Sterkfontein Caves: Official Visitor Centres for the Cradle of Humankind World Heritage Site, May 20, 2019, https://www .maropeng.co.za/news/entry/celebrate-Africa-day-celebrate-humanity.

25. David Begun, Mariam C. Nargolwalla, and László Kordos, "European Miocene Hominids and the Origin of the African Ape and Human Clade," *Evolutionary Anthropology* 21, no. 1 (2012): 10–23.

26. Darwin, *Descent*, 1:199.

27. John Edward Gray, "An Outline of an Attempt at the Disposition of Mammalia into Tribes and Families, with a List of the Genera Apparently Appertaining to Each Tribe," *Annals of Philosophy* 10 (1825): 337–44.

28. Mivart, "Axial Skeleton Primates"; Mivart, "Appendicular Skeleton of the Primates."

29. Thomas Henry Huxley, *An Introduction to the Classification of Animals* (London: J. Churchill & Sons, 1869).

30. Lyell, *Antiquity of Man*.

31. Johann F. Blumenbach, *Handbuch der Naturgeschichte* (Göttingen: J. C. Dieterich, 1779); Georges Cuvier, *Le Règne Animal Distribué d'Après son Organisation, pour Servir de Base à l'Histoire Naturelle des Animaux et d'Introduction à l'Anatomie Comparèe* (Paris: Imprimerie de A. Belin, 1816).

32. Haeckel, *Generelle Morphologie*; Ernst Haeckel, *Natürliche Schöpfungsgeschichte: Gemeinverständliche wissenschaftliche Vorträge über die Entwickelungslehre im Allgemeinen und diejenige von Darwin, Göthe und Lamarck im Besonderen, über die Anwendung derselben auf den Ursprung des Menschen und andern damit zusammenhängende Gründfragen der Natur-Wissenschaft. Mit Tafeln, Holzschnitten, systematischen und genealogischen Tabellen* (Jena: Walter de Gruyter, 1868); Étienne Geoffroy Saint-Hilaire, *Tableau des Quadrumanes, ou des Animaux*

Composant le Premier Ordre de la Classe des Mammifères (Paris: Muséum National d'Histoire Naturelle, 1812).

33. George Gaylord Simpson, "The Principles of Classification and a Classification of Mammals," *Bulletin of the American Museum of Natural History* 85 (1945), 187.

34. Philip D. Cantino and Kevin de Queiroz, "PhyloCode: A Phylogenetic Code of Biological Nomenclature" (unpublished manuscript, 2000).

35. Darwin, *Descent*, 1:201.

36. Emile Zuckerkandl, Richard T. Jones, and Linus Pauling, "A Comparison of Animal Hemoglobins by Tryptic Peptide Pattern Analysis," *Proceedings of the National Academy of Sciences* 46, no. 10 (1960): 1349; Morris Goodman, "Immunochemistry of the Primates and Primate Evolution," *Annals of the New York Academy of Sciences* 102, no. 2 (1962): 219–34.

37. Colin P. Groves, *A Theory of Human and Primate Evolution* (Oxford: Clarendon Press, 1989).

38. See, for example, Tim D. White et al., "*Ardipithecus ramidus* and the Paleobiology of Early Hominids," *Science* 326, no. 5949 (2009): 64–86.

39. Simpson, "Principles of Classification," 187.

40. Charles Hamilton Smith, *The Natural History of the Human Species: Its Typical Forms, Primeval Distribution, Filiations, and Migrations* (London: Gould and Lincoln, 1855).

41. Richard Owen, *On the Classification and Geographical Distribution of the Mammalia: Being the Lecture on Sir Robert Reade's Foundation, Delivered before the University of Cambridge in the Senate House, May 10, 1859; To Which Is Added an Appendix "On the Gorilla," and "On the Extinction and Transmutation of Species"* (London: John W. Parker and Son, 1859).

42. Darwin, *Descent*, 1:201.

43. Alfred Russel Wallace, "The Origin of Human Races and the Antiquity of Man Deduced from the Theory of 'Natural Selection,'" *Journal of the Anthropological Society of London* 2 (1864), clviii–clxxxvii.

44. Charles Darwin, *On the Origin of Species by Means of Natural Selection, or the Preservation of Favoured Races in the Struggle for Life* (London: John Murray, 1859).

45. Darwin, *Descent*, 1:188.

46. Jan Ole Kriegs, Gennady Churakov, Jerzy Jurka, Jürgen Brosius, and Jürgen Schmitz, "Evolutionary History of 7SL RNA-Derived SINEs in Supraprimates," *Trends in Genetics* 23, no. 4 (2007): 158–61.

47. Darwin, *Descent*, 1:213.

Chapter 7: "On the Races of Man": Race, Racism, Science, and Hope

1. Charles Darwin, *The Descent of Man, and Selection in Relation to Sex*, 1st ed. (London: John Murray, 1871), 1:214.

2. Ibid., 215.

3. Ibid., 215–16.

4. Ibid., 216.

5. Ibid., 219, 221, 225–26.

6. Ibid., 232. By "intimate" Darwin simply means that he knew, interacted and learned from the individual. Some suggest that Edmonstone was influential in interesting Darwin in embarking on travels to South America. See Figure 7.1.

7. Ibid., 231–32.

8. Jon Marks, *Is Science Racist?* (Cambridge: Polity, 2017); Angela Saini, *Superior: The Return of Race Science* (Boston: Beacon Press, 2019).

9. Darwin, *Descent*, 1:241–42.

10. Ibid., 246.

11. Ibid., 247.

12. Ibid., 248–49.

13. Ibid., 249.

14. Ibid.

15. Ibid., 236.

16. Ibid., 238.

17. Ibid.

18. Ibid., 239.

19. Ibid., 238.

20. Ibid.

21. Charles Darwin, *The Descent of Man, and Selection in Relation to Sex*, 2nd ed. (London: John Murray, 1874), 190. See whole text section associated with this passage and footnote 47.

22. Ibid., 239–40.

23. Agustín Fuentes et al., "AAPA Statement on Race and Racism," *American Journal of Physical Anthropology* 169 (2019): 400–402, https://doi.org/10.1002/ajpa.23882; Agustín Fuentes, *Race, Monogamy and Other Lies They Told You: Busting Myths about Human Nature* (Berkeley: University of California Press, 2012); Keith L. Hunley, Graciela S. Cabana, and Jeffrey C. Long, "The Apportionment of Human Diversity Revisited," *American Journal of Physical Anthropology* 160 (2016): 561–69; Jon Marks, "Ten Facts about Human Variation," in *Human Evolutionary Biology*, ed. Michael P. Muehlenbein (Cambridge: Cambridge University Press, 2010), 265–76; Alan Templeton, "Biological Races in Humans," *Studies in History and Philosophy of Biological and Biomedical Sciences* 44 (2013): 262–71.

24. Seela Athreya, "Picking a Bone with Evolutionary Essentialism," *Anthropology News* 59, no. 5 (2018): e55–e60; Adam Van Arsdale, "Population Demography, Ancestry, and the Biological Concept of Race," *Annual Review of Anthropology* 48 (2019), 227–41; Sang Hee Lee, "Where Do We Come From?" *Anthropology News* 59, no. 4 (2018): 18–20.

25. Fuentes, "Statement on Race," 400–402.

26. American Society of Human Genetics, "ASHG Denounces Attempts to Link Genetics and Racial Supremacy," *American Journal of Human Genetics* 103, no. 5 (2018), 636.

27. Deborah A. Bolnick, "Individual Ancestry Inference and the Reification of Race as a Biological Phenomenon," in *Revisiting Race in a Genomic Age*, eds. Barbara A. Koenig, Sandra Lee, and Sarah S. Richardson (New Brunswick, NJ: Rutgers University Press, 2008): 70–88; Joseph Graves Jr., "Why the Nonexistence of Biological Races Does Not Mean the Nonexistence of Racism," *American Behavioral Scientist* 59 (2015): 1474–95; Hunley et al., "Apportionment of

Human Diversity," 561–69; Templeton, "Biological Races in Humans," 262–71; A. Bergström et al., *Science* 367 (2020): eaay5012, doi:10.1126/science.aay5012.

28. Fuentes, "Statement on Race," 400–402; Fuentes, *Race, and Other Lies*; Marks, "Ten Facts," 265–76.

29. Alondra Nelson, *The Social Life of DNA: Race, Reparations, and Reconciliation after the Genome* (Boston: Beacon Press, 2016); Dorothy Roberts, *Fatal Invention: How Science, Politics, and Big Business Re-Create Race in the Twenty-First Century* (New York: The New Press, 2012); Saini, *Superior*.

30. Graves, "Nonexistence of Biological Races," 1474–95; Marks, "Ten Facts," 265–276; Nelson, *Social Life of DNA*; Roberts, *Fatal Invention*; Saini, *Superior*.

31. Fuentes, *Race, and Other Lies*; Clarence C. Gravlee, Amy Non, and Connie Mulligan, "Genetic Ancestry, Social Classification, and Racial Inequalities in Blood Pressure in Southeastern Puerto Rico," *PLoS One* 4, no. 9 (2009): e6821; Jacklyn Quinlan et al., "Genetic Loci and Novel Discrimination Measures Associated with Blood Pressure Variation in African Americans Living in Tallahassee," *PLoS One* 11, no. 12 (2016): e0167700; Nancy Krieger, "Measures of Racism, Sexism, Heterosexism, and Gender Binarism for Health Equity Research: From Structural Injustice to Embodied Harm—an Ecosocial Analysis," *Annual Review of Public Health* 41 (2020): 1–26.

32. For example, Nicholas Wade, *A Troublesome Inheritance: Genes, Race and Human History* (New York: Penguin Press, 2014); Richard J. Herrnstein and Charles Murray, *The Bell Curve: Intelligence and Class Structure in American Life* (New York: Free Press, 1994); Charles Murray, *Human Diversity: The Biology of Gender, Race, and Class* (New York: Twelve, 2020).

33. Darwin, *Descent*, 1st ed., 1:249.

34. Steven Rose, "Darwin, Race and Gender," *European Molecular Biology Organization Reports* 10, no. 4 (2009): 297–298.

35. Adrian Desmond and James Moore, *Darwin's Sacred Cause: How a Hatred of Slavery Shaped Darwin's Views on Human Evolution* (Boston & New York: Houghton Mifflin Harcourt, 2009).

36. Fuentes, *Race, and Other Lies*; Agustín Fuentes, *Why We Believe: Evolution and the Human Way of Being* (New Haven, CT: Yale University Press, 2019).

Chapter 8: Resolving the Problem of Sexual Beauty

1. Darwin to Asa Gray, April 3, 1860, Darwin Correspondence Project, letter no. 2743, https://www.darwinproject.ac.uk/letter/DCP-LETT-2743.xml.

2. Fabienne Smith, "Charles Darwin's Ill Health," *Journal of the History of Biology* 23 (1990): 443–59.

3. Peter R. Grant and B. Rosemary Grant, *How and Why Species Multiply: The Radiation of Darwin's Finches* (Princeton, NJ: Princeton University Press, 2011).

4. Arhat Abzhanov et al., "Bmp4 and Morphological Variation of Beaks in Darwin's Finches," *Science* 305 (2004): 1462; Arhat Abzhanov et al., "The Calmodulin Pathway and Evolution of Elongated Beak Morphology in Darwin's Finches," *Nature* 442 (2006): 563–67.

5. Charles Darwin, *On the Origin of Species by Means of Natural Selection, or the Preservation of Favoured Races in the Struggle for Life*, 1st ed. (London: John Murray, 1859).

6. Thomas R. Malthus, *An Essay on the Principle of Population; or, a View of Its Past and Present Effects on Human Happiness* (London: Reeves & Turner, 1888).

7. See, for example, John A. Endler, *Natural Selection in the Wild* (Princeton, NJ: Princeton University Press, 1986).

8. Kazunori Yoshizawa, Rodrigo Ferreira, Yoshitaka Kamimura, and Charles Lienhard, "Female Penis, Male Vagina, and Their Correlated Evolution in a Cave Insect," *Current Biology* 24 (2014): 1006–10.

9. Robert L. Trivers, "Parental Investment and Sexual Selection," in *Sexual Selection and the Descent of Man*, ed. Bernard G. Campbell (Chicago: Aldine, 1972), 136–79.

10. John Maynard Smith, *The Evolution of Sex* (Cambridge: Cambridge University Press, 1978).

11. Ibid.

12. Angus J. Bateman, "Intrasexual Selection in *Drosophila*," *Heredity* 2 (1948): 349–68.

13. Zuleyma Tang-Martínez, "Rethinking Bateman's Principles: Challenging Persistent Myths of Sexually Reluctant Females and Promiscuous Males," *Journal of Sex Research* 53 (2016): 532–59; Angela Saini, *Inferior: How Science Got Women Wrong—and the New Research That's Rewriting the Story* (Boston: Beacon Press, 2017).

14. Adam G. Jones et al., "The Bateman Gradient and the Cause of Sexual Selection in a Sex-Role-Reversed Pipefish," *Proceedings of the Royal Society of London, ser. B: Biological Sciences* 267 (2000): 677–80.

15. Michael J. Wade and Stephen M. Shuster, "Don't Throw Bateman Out with the Bathwater!" *Integrative and Comparative Biology* 45 (2005): 945–51.

16. Stephen T. Emlen and Lewis W. Oring, "Ecology, Sexual Selection, and the Evolution of Mating Strategies," *Science* 197 (1977): 215–23.

17. Douglas J. Emlen, *Animal Weapons: The Evolution of Battle* (New York: Henry Holt, 2014).

18. R. Haven Wiley, "Territoriality and Non-Random Mating in Sage Grouse, *Centrocercus urophasianus*," *Animal Behaviour Monographs* 6 (1973): 85–169.

19. David M. Buss, *The Evolution of Desire* (New York: Basic Books, 1994).

20. Mart R. Gross, "Sunfish, Salmon, and the Evolution of Alternative Reproductive Strategies and Tactics in Fishes," *Fish Reproduction* (1984): 55–75.

21. Alfred R. Wallace, "Lessons from Nature, as Manifested in Mind and Matter," *Academy* 562 (1876).

22. Alfred Tennyson, *In Memoriam AHH* (Los Angeles: Hardpress Publishing, 2012 [1850]).

23. Evelleen Richards, *Darwin and the Making of Sexual Selection* (Chicago: University of Chicago Press, 2017).

24. Malte Andersson, *Sexual Selection* (Princeton, NJ: Princeton University Press, 1994); Richard O. Prum, *The Evolution of Beauty: How Darwin's Forgotten Theory of Mate Choice Shapes the Animal World—and Us* (New York: Doubleday, 2017); Gil G. Rosenthal, *Mate Choice: The Evolution of Sexual Decision Making from Microbes to Humans* (Princeton, NJ: Princeton University Press, 2017); Michael J. Ryan, *A Taste for the Beautiful: The Evolution of Attraction* (Princeton, NJ: Princeton University Press, 2018).

25. Ryan, *A Taste for the Beautiful*.

26. Rosenthal, *Mate Choice*; Jerry A. Coyne and H. Allen Orr, *Speciation* (Sunderland, MA: Sinauer, 2004).

27. Marion Petrie, "Improved Growth and Survival of Offspring of Peacocks with More Elaborate Trains," *Nature* 371 (1994): 598–99; Allison M. Welch, Raymond D. Semlitsch, and H. Carl Gerhardt, "Call Duration as an Indicator of Genetic Quality in Male Gray Tree Frogs," *Science* 280 (1998): 1928–30.

28. Gerald S. Wilkinson and Paul R. Reillo, "Female Choice Response to Artificial Selection on an Exaggerated Male Trait in a Stalk-Eyed Fly," *Proceedings of the Royal Society of London, ser. B: Biological Sciences* 255 (1994): 1–6.

29. F. Helen Rodd, Kimberly A. Hughes, and Trevor E. Pitcher, "Sex, Color and Mate Choice in Guppies," *Reviews in Fish Biology and Fisheries* 9 (1999): 203–7.

30. Rosenthal, *Mate Choice*; Hanna Kokko, Robert Brooks, Michael D. Jennions, and Josephine Morley, "The Evolution of Mate Choice and Mating Biases," *Proceedings of the Royal Society of London, ser. B: Biological Sciences* 270 (2003): 653–64.

31. Reviewed in Thomas A. Sebeok, *The Sign and Its Masters* (Austin: University of Texas Press, 1979).

32. Michael J. Ryan and Molly E. Cummings, "Perceptual Biases and Mate Choice," *Annual Review of Ecology, Evolution, and Systematics* 44 (2013): 437–59.

33. Molly E. Cummings, "Sensory Trade-Offs Predict Signal Divergence in Surfperch," *Evolution* 61 (2007): 530–45.

34. Walter Wilczynski, Ann Keddy-Hector, and Michael J. Ryan, "Patterns and Basilar Papilla Tuning in Cricket Frogs, I. Differences among Populations and between Sexes," *Brain, Behavior & Evolution* 39 (1992): 229–37.

35. Kim L. Hoke et al., "Functional Mapping of the Auditory Midbrain during Mate Call Reception," *Journal of Neuroscience* 24 (2004): 11264–72; Michael J. Ryan, *The Túngara Frog: A Study in Sexual Selection and Communication* (Chicago: University of Chicago Press, 1985).

36. Roger N. Shepard, "Toward a Universal Law of Generalization for Psychological Science," *Science* 237 (1987): 1317–23.

37. Karin L. Akre et al., "Signal Perception in Frogs and Bats and the Evolution of Mating Signals," *Science* 333 (2011): 751–52.

38. Jeffrey Gassen et al., "Beauty Is in the Psychophysics of the Beholder: Facial Attractiveness, Sexual Selection, and Weber's Law," (unpublished manuscript, 2018).

39. James W. Pennebaker et al., "Don't the Girls Get Prettier at Closing Time: A Country and Western Application to Psychology," *Personality and Social Psychology Bulletin* 5 (1979): 122.

40. Carly Johnco, Ladd Wheeler, and Alan J. Taylor, "They Do Get Prettier at Closing Time: A Repeated Measures Study of the Closing-Time Effect and Alcohol," *Social Influence* 5 (2010): 261–71.

41. Kathleen S. Lynch, A. Stanley Rand, Michael J. Ryan, and Walter Wilczynski, "Reproductive State Influences Female Plasticity in Mate Choice," *Animal Behaviour* 69 (2005): 689–99.

42. Ingo Schlupp, Catherine A. Marler, and Michael J. Ryan, "Benefit to Male Sailfin Mollies of Mating with Heterospecific Females," *Science* 263 (1994): 373–74.

43. Sarah E. Hill and Michael J. Ryan, "The Role of Model Female Quality in the Mate Choice Copying Behaviour of Sailfin Mollies," *Biology Letters* 2 (2006): 203–5.

44. Amany Gouda-Vossos, Shinichi Nakagawa, Barnaby J. W. Dixson, and Robert C. Brooks, "Mate Choice Copying in Humans: A Systematic Review and Meta-Analysis," *Adaptive Human Behavior and Physiology* 4 (2018): 364–86.

45. Sally E. Street et al., "Human Mate-Choice Copying Is Domain-General Social Learning," *Scientific Reports* 8 (2018): 1715.

46. Charles Darwin, *The Descent of Man, and Selection in Relation to Sex*, 1st ed. (London: John Murray, 1871), 1:64.

Chapter 9: This View of Wife

1. Charles Darwin, *The Descent of Man, and Selection in Relation to Sex*, 1st ed. (London: John Murray, 1871), 2:316.

2. Charles Darwin, *The Descent of Man, and Selection in Relation to Sex*, 2nd ed. (London: John Murray, 1874), 563. In the first edition, the phrase "in securing wives" is used instead of "in their contests for wives."

3. Darwin, *Descent*, 1st ed., 2:326–27.

4. Ibid., 329.

5. Ibid., 385.

6. Darwin, *Descent*, 2nd ed., 556.

7. Ibid., 558–59.

8. Nina Jablonski, "Skin Color," in *The International Encyclopedia of Biological Anthropology*, ed. Wenda Trevathan (New York: Wiley & Sons, 2018), 1430–34.

9. Nina Jablonski and George Chaplin, "The Evolution of Human Skin Coloration," *Journal of Human Evolution* 39 (2000): 57–106.

10. Jablonski, "Skin Color."

11. Ellen Quillen et al., "Shades of Complexity: New Perspectives on the Evolution and Genetic Architecture of Human Skin," *American Journal of Physical Anthropology* 168 (2018): 4.

12. Tina Lasisi and Mark D. Shriver, "Focus on African Diversity Confirms Complexity of Skin Pigmentation Genetics," *Genome Biology* 19 (2018): 13.

13. Jablonski, "Skin Color," 1433.

14. Paolo U. Giacomoni, Thomas Mammone, and Matthew Teri, "Gender-Linked Differences in Human Skin," *Journal of Dermatological Science* 55 (2009): 144–49.

15. Lorena Madrigal and William Kelly, "Human Skin-Color Sexual Dimorphism: A Test of the Sexual Selection Hypothesis," *American Journal of Physical Anthropology* 132 (2007): 470–82.

16. Smith (2016) referred to this logical trap as "current utility versus historical role" in his review of this and related issues in the science of paleoanthropology. See Richard J. Smith, "Explanations for Adaptations, Just-so Stories, and Limitations on Evidence in Evolutionary Biology," *Evolutionary Anthropology* 25 (2016): 276–87.

17. Darwin, *Descent*, 2nd ed., 563.

18. Holly M. Dunsworth, "Expanding the Evolutionary Explanations for Sex Differences in the Human Skeleton," *Evolutionary Anthropology* (May 2020): 108–116, https://doi.org/10.1002/evan.21834.

19. Giacomoni, "Gender-Linked Differences," 144–49.

20. Sue T. Parker and Karen E. Jaffe, *Darwin's Legacy: Scenarios in Human Evolution* (Plymouth, UK: Altamira, 2008).

21. Ng Leo and Stephen Barker, "Unravelling the Evolution of the Head Lice and Body Lice of Humans," *Parasitology Research* 98, no. 1 (2005): 44–47.

22. Darwin, *Descent*, 1st ed., 2:376.

23. Peter F. Wheeler, "The Evolution of Bipedality and Functional Body Hair in Hominids," *Journal of Human Evolution* 13 (1984): 91–98; Dean Falk, "Brain Evolution in *Homo*: The 'Radiator' Theory," *Behavioral and Brain Sciences* 13 (1990): 333–81

24. Darwin to Alfred Russel Wallace, December 22, 1857, Darwin Correspondence Project, letter no. 2192, https://www.darwinproject.ac.uk/letter/DCP-LETT-2192.xml.

25. Alison Gopnik, *The Scientist in the Crib* (New York: Perennial, 2001), 201.

26. Ibid., 85, my emphasis.

27. Darwin, *Descent*, 1st ed., 2:327.

28. Thomas H. Huxley, *Evidence as to Man's Place in Nature* (London: Williams and Norgate, 1863), 69.

29. Edward O. Wilson, *Sociobiology* (Cambridge, MA: Harvard University Press, 1975), 547.

30. Anne Fausto-Sterling, *Myths of Gender: Biological Theories about Women and Men*, 2nd ed. (New York: Basic Books, 1992).

31. Rick W. A. Smith and Deborah A. Bolnick, "Situating Science: Doing Biological Anthropology as a View from Somewhere," *American Anthropology* 121, no. 2 (2019): 465–67.

32. Robin G. Nelson, "Hypervisible and Human," *American Anthropologist* 121, no. 2 (2019): 469–70.

33. Kim Tallbear, "Feminist, Queer, and Indigenous Thinking as an Antidote to Masculinist Objectivity and Binary Thinking in Biological Anthropology," *American Anthropologist* 121, no. 2 (2019): 494–96.

34. Darwin to Alfred Russel Wallace, December 22, 1857.

35. Lynn Margulis, *Symbiosis in Cell Evolution: Life and Its Environment on the Early Earth* (Boston: WH Freeman, 1981).

36. Kenneth M. Weiss and Anne V. Buchanan, *The Mermaid's Tale: Four Billion Years of Cooperation in the Making of Living Things* (Cambridge, MA: Harvard University Press, 2009).

37. Rebecca J. Lewis, "Female Power in Primates and the Phenomenon of Female Dominance," *Annual Review of Anthropology* 47 (2018): 542.

38. Sarah Blaffer Hrdy, *The Woman That Never Evolved: With a New Preface and Bibliographical Updates*, rev. ed. (Cambridge, MA: Harvard University Press, 1999), 13.

39. For more on this, see Angela Saini, *Inferior: How Science Got Women Wrong and the New Research That's Rewriting the Story* (Boston: Beacon Press, 2017), 108.

40. Barbara Smuts, "Evolutionary Origins of Patriarchy," *Human Nature* 6, no. 1 (1995): 1–32.

41. Lewis, "Female Power," 539, emphasis mine.

42. Ibid., 541, emphasis mine.

43. Saini, *Inferior*, 150.

44. Darwin, *Descent*, 1st ed., 2:320.

45. Hrdy, *Woman That Never Evolved*, 59.

46. Kimberly A. Hamlin, *From Eve to Evolution: Darwin, Science and Women's Rights in Gilded Age America* (Chicago: University of Chicago Press, 2014), 70–73.

47. Venla Oikkonen, *Gender, Sexuality and Reproduction in Evolutionary Narratives* (New York: Routledge, 2013).

48. Cynthia Taylor and Bryan M. Dewsbury, "On the Problem and Promise of Metaphor Use in Science and Science Communication," *Journal of Microbiology and Biology Education* 19, no. 1 (2018): 1–5.

49. Misia Landau, *Narratives of Human Evolution* (New Haven, CT: Yale University Press, 1991).

50. Fausto-Sterling, *Myths of Gender*, 183.

51. Wednesday Martin, *Untrue: Why Nearly Everything We Believe about Women, Lust, and Infidelity Is Wrong and How the New Science Can Set Us Free* (New York: Little Brown, 2018), 167.

52. Martin, *Untrue*, 45.

53. Charles Darwin, *On the Origin of Species by Means of Natural Selection, or the Preservation of Favoured Races in the Struggle for Life*, 1st ed. (London: John Murray, 1859).

Chapter 10: Dinner with Darwin: Sharing the Evidence Bearing on the Origin of Humans

1. Charles Darwin, *The Descent of Man, and Selection in Relation to Sex*, 1st ed. (London: John Murray, 1871), 2:405.

2. Tori Reeve, *Down House: The Home of Charles Darwin* (London: English Heritage, 2009), 34.

3. Ibid., 20.

4. Ibid., 24.

5. Charles Darwin, *Insectivorous Plants* (London: John Murray, 1875), 286.

6. Ibid., 289.

7. Reeve, *Down House*, 25.

8. Janet Browne, *Darwin: The Power of Place* (Princeton, NJ: Princeton University Press, 2002), 343.

9. Charles Darwin, *The Descent of Man: The Concise Edition*, ed. Carl Zimmer (New York: Plume, 2007), 1.

10. Alex Menez, "The Gibraltar Skull: Early History, 1848–1868," *Archives of Natural History* 45 (2018): 92–110.

11. Darwin to J. D. Hooker, September 1, 1864, Darwin Correspondence Project, letter no. 4605, https://www.darwinproject.ac.uk/letter/DCP-LETT-4605.xml.

12. Michael Balter, "When Darwin Met a Neanderthal," *Science* online, September 22, 2009, https://blogs.sciencemag.org/origins/2009/09/when-darwin-met-a-neandertal.html.

13. Dusha Bateson, *Mrs. Charles Darwin's Recipe Book* (New York: Glitterati, 2012).

14. Darwin, *Descent*, 1st ed., 1:199.

15. Browne, *Power of Place*, 343.

16. Chris Stringer, e-mail message to author, May 7, 2019.

17. Darwin, *Descent*, 1st ed., 1:156–57.

18. Ibid., 199.

19. Adrienne Zihlman, e-mail message to author, June 5, 2019.

20. Kevin Langergraber et al., "Generation Times in Wild Chimpanzees and Gorillas Suggest Earlier Divergence Times in Great Ape and Human Evolution," *Proceedings of the National Academy of Sciences* 109 (2012): 15716–21.

21. Adrienne Zihlman, "Pygmy Chimps, People, and the Pundits," *New Scientist* 104 (1984): 39–40.

22. Leslie Aiello, e-mail message to author, June 7, 2019.

23. Francis Thackeray, e-mail message to author, May 8, 2019.

24. Ibid.

25. Mary Leakey, *Disclosing the Past: An Autobiography* (New York: Doubleday, 1984), 120–21.

26. Virginia Morell, *Ancestral Passions: The Leakey Family and the Quest for Humankind's Beginnings* (New York: Simon & Schuster, 1995).

27. Charles Darwin, *On the Origin of Species by Means of Natural Selection, or the Preservation of Favoured Races in the Struggle for Life*, 1st ed. (London: John Murray, 1859), 759.

28. William Glen, *The Road to Jaramillo: Critical Years of the Revolution in Earth Sciences* (Stanford, CA: Stanford University Press, 1982).

29. Ibid., 77.

30. Tim D. White, from CD of talk at the 2003 Nobel Conference at Gustavus Adolphus College, Saint Peter, MN, October 7, 2003.

31. Darwin, *Descent*, 1st ed., 1:126.

32. Ann Gibbons, "A New Kind of Ancestor: *Ardipithecus* Unveiled," *Science* 326 (2009): 36.

33. Ann Gibbons, "A Human Smile and Funny Walk for *Australopithecus sediba*," *Science* 340 (2013): 132.

34. Ann Gibbons, "Evolution's Case Evolves," *Los Angeles Times*, April 22, 2006.

35. Ann Gibbons, "The Wanderers," *Science* 354 (2016): 958.

36. Lizzie Wade, "New Species of Ancient Human Unearthed," *Science* 364 (2019): 108.

37. Ann Gibbons, "First Modern Humans in China," *Science* 350 (2015): 264.

38. Ann Gibbons, "Oldest Members of Our Species Discovered in Morocco," *Science* 356 (2017): 993.

39. Lizzie Wade, "Was Our Species in Europe 210,000 Years Ago?" *Science* 365 (2019): 111.

40. Chris Stringer, "When Did Modern Humans Leave Africa?" *Science* 359 (2018): 389.

41. Viviane Slon, e-mail message to author, June 15, 2019.

42. Fahu Chen et al., "A Late Middle Pleistocene Denisovan Mandible from the Tibetan Plateau," *Nature* 569 (2019): 409–12.

43. Ann Gibbons, "Ancient Jaw Gives Elusive Denisovans a Face," *Science* 364 (2019): 418–19.

44. Chris Stringer, e-mail message to author, May 7, 2019.

45. Viviane Slon et al., "Neandertal and Denisovan DNA from Pleistocene Sediments," *Science* 356 (2017): 605.

46. Ann Gibbons, "Spotting Evolution among Us," *Science* 363 (2019): 21.

47. Darwin, *Descent*, 1st ed., 1:13.

INDEX

Index Note: Illustrations are indicated with **bold** numbers.

A NOTE ON THE TYPE

This book has been composed in Arno, an Old-style serif typeface in the classic Venetian tradition, designed by Robert Slimbach at Adobe.